Physics and Chemistry in Space
Volume 8

Edited by
J. G. Roederer, Denver
J. T. Wasson, Los Angeles

Akira Hasegawa

Plasma Instabilities and Nonlinear Effects

With 48 Figures

Springer-Verlag
Berlin Heidelberg New York 1975

AKIRA HASEGAWA
Bell Telephone Laboratories, Murray Hill, NJ 07974/USA

The illustration on the cover is adapted from Fig. 30, which shows various shapes of instabilities of cylindrical current pinches.

ISBN 3-540-06947-X Springer Verlag Berlin Heidelberg New York
ISBN 0-387-06947-X Springer-Verlag New York Heidelberg Berlin

Library of Congress Cataloging in Publication Data. Hasegawa, Akira, 1934—. Plasma instabilities and nonlinear effects. (Physics and chemistry in space, 8). Bibliography: p. 1. Plasma instabilities. I. Title. II. Series. QC801.P46 vol 8 [QC718.5.S7] 523.01'08s [530.4'4]. 74—18072

Typesetting, printing and bookbinding: Universitätsdruckerei H. Stürtz AG, Würzburg

To my mother
Kaoru Takata

Preface

In recent years the significant progress in satellite-based observations of plasma states and associated electromagnetic phenomena in space has resulted in the accumulation of much evidence of various plasma instabilities. Today plasma instabilities are believed to be responsible for electromagnetic radiation as well as for many of the macroscopic dynamics of plasmas in space.

Most students who begin to study plasma physics are intrigued by the unstable nature of plasmas compared with other states of matter; however, they often become frustrated because there are so many instabilities. Such frustration explains in part why there is no textbook which treats this subject exclusively. A description of plasma instabilities in a systematic way is nontrivial and takes a pertinacious effort. This book is an attempt to provide a basic introduction on the subject and covers most of the important instabilities. However, the author must apologize for any omission of references to contributions of individuals who deserve more credit.

The reader is assumed to have a general knowledge of plasma physics obtainable in an undergraduate course. The book is intended to be used as a reference text on the subject of plasma instabilities at the undergraduate level as well as for a text in a special course in graduate school. Because the book is part of a series on physics and chemistry in space, emphasis is placed on plasma instabilities relevant in space plasmas. However, most of the instabilities discussed here are also applicable to laboratory plasmas; hence the book should serve also as an introductory text for scientists interested in controlled thermonuclear fusion. In fact, most of the instabilities introduced in the text have been discovered in the course of research in laboratory plasmas.

The book is divided into four chapters. The first is an introduction to plasma instabilities, where their general concept and physical origin are presented. They are classified in standard form as microscopic instabilities (those associated with a velocity space nonequilibrium) and macroscopic instabilities (those associated with a coordinate space nonequilibrium). These are described in the second and the third chapters. Following the theoretical derivations, the implications of the instabilities

and possible satellite observations are presented for as many cases as possible. To consider some possible consequences of the instabilities, associated nonlinear processes are discussed in chapter 4.
MKS units are used throughout the book.

The author wishes to thank Mr. T-Y Huang for his careful checking of algebra, Ms. C. G. Maclennan for her editing of English and Mr. B. A. Stevens for collection of references, as also to express his gratitude to Dr. H. L. Berk, Dr. L. Chen, Prof. H. P. Furth, Dr. L. J. Lanzerotti, Dr. L. M. Linson, Prof. T. M. O'Neil, Prof. T. Taniuti and Dr. F. D. Tappert for their valuable discussions and many important suggestions.

The book has grown in part from the lecture notes for graduate courses in Columbia University Plasma Physics Group by the author in spring terms of 1972 and 1974, and in part from the author's review article "Plasma Instabilities in the Magnetosphere", Review of Geophysics and Space Physics **9**, 703 (1971). Discussions with the members of the Columbia Plasma Physics Group, in particular Profs. C. K. Chu, T. C. Marshall and A. K. Sen have been very valuable.

The author wishes to extend his appreciation for the encouragement given by Dr. W. L. Brown and Dr. J. A. Giordmaine of Bell Laboratories, Prof. C. K. Birdsall of the University of California, Prof. A. J. Dessler of Rice University and Prof. J. G. Roederer of the University of Denver.

The author is indebted to Mrs. E. H. Nitchie, Miss D. G. Bracht and many typists in Bell Laboratories typing pool for their careful typing of the manuscript.

Finally the author is grateful to his wife Miyoko for her encouragement to the project and her gracious hospitality to my companions, Johann Sebastian Bach, Miles Davis, A. Stockly, and Sachiko Nishida (and some Scotch) during the preparation of the manuscript.

New Jersey, 1975

Akira Hasegawa

Contents

1. Introduction to Plasma Instabilities

1.1 Introduction

Particles in space plasmas generally have a large mean free distance between interparticle collisions. For example, the mean free distance of solar wind electrons is known to be approximately 1 A. U. near the earth. In terms of the mean free time, this means that such a particle will collide only once every few days. Furthermore, if we note that the energy equipartition time is longer than the interparticle collision time by the mass ratio of proton and electron, we can expect that the plasmas in space, either interplanetary or magnetospheric, are not in thermodynamical equilibrium. However, these plasmas can in most cases be assumed to be under dynamical equilibrium: forces acting on the plasma body are balanced to zero. This is because the time scale of a dynamic response of a plasma will be $\sim\omega_p^{-1}$ or ω_c^{-1} where ω_p and ω_c are plasma and cyclotron (angular) frequencies respectively. For the example of the solar wind these numbers are approximately 10^{-5} and 10^{-3} sec for electrons and 10^{-2} and 1 sec for protons.

The fact that a plasma is not in thermodynamic equilibrium implies that a certain amount of free energy is stored in the plasma; this energy may be converted into a violent motion of the plasma or into radiation of electromagnetic waves. Plasma instability is a process where such a conversion takes place in a collective way: this leads to the fact that a small deviation from the dynamic equilibrium becomes the cause of a further deviation. Mathematically, if we write such a deviation x, the above statement means that the time rate change of x, dx/dt, is proportional to x itself, say, $dx/dt=\gamma x$. Solving this equation, we obtain $x=e^{\gamma t}$, showing that the deviation grows exponentially. The constant γ is called the growth rate of the instability.

There are basically two ways by which a plasma can deviate from thermodynamic equilibrium. One way is by localizing in space with a locally higher (or lower) density, temperature, pressure or some other thermodynamic quantity; the other is by having a velocity distribution other than the Maxwell-Boltzmann distribution. When an instability occurs because of the former reason, the plasma as a whole will change

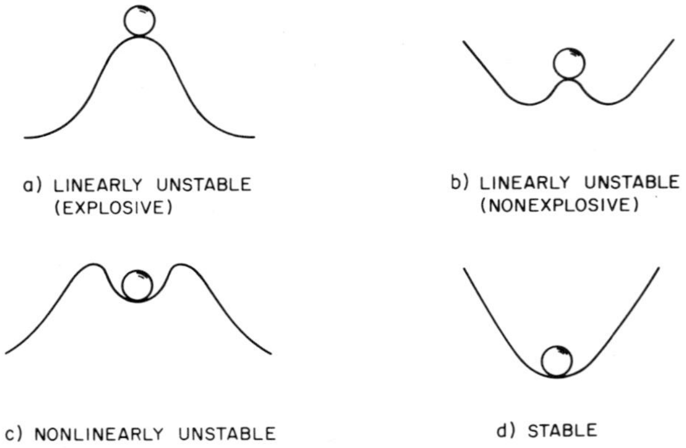

Fig. 1a–d. Models of various stability conditions

its shape, an external process, hence called a macroscopic instability. An instability caused by the latter, an internal process, is called a microscopic instability.

Plasma instability is often demonstrated by a ball sitting on top of a hill. We will use such a model to explain various different ways in which an instability can develop. Commonly an unstable state and a stable state are represented as in Fig. 1a and d. Note that the ball in either case is in dynamical equilibrium.

Fig. 1a represents a linearly unstable situation because an infinitely small perturbation could knock the ball down. Case a) may be called explosively unstable because a finite displacement of the ball position does not lead to a stable situation, whereas, in contrast, in case b) the ball reaches the next hill and is reflected back, thus it will be stabilized with a finite (nonzero) size of perturbation (displacement). Case c), on the other hand, may look stable, but if a large enough perturbation is applied to the ball, it becomes unstable. Such a case is generally called nonlinearly unstable. Case d) represents an absolute stability.

For careful readers, it is well to point out that this is just a model, and strictly speaking, Fig. 1a *does not* represent a linear instability. To be a linear instability, the time rate change of the displacement of the ball position must increase, by definition, in proportion to the amount of displacement itself. Therefore one must take into account the change of gravity as a function of the displacement.

An explosive instability is defined as one in which the growth rate γ is an increasing function of time; in other words, the time rate change of the displacement increases with a further increase of the displacement.

I leave readers to judge under what conditions Fig. 1a correctly represents an explosive situation. A more detailed discussion of an explosive instability will be given in Chapter 4.

Let us now ask why plasma instability is important or interesting in space physics. As can be seen from the foregoing argument, a plasma through an instability can discharge its free energy and will try to reach thermodynamical equilibrium. The rate of such a change is determined ordinarily by the dynamic time scale which is much faster than the rate due to collisional process. This way a plasma instability by its collective effect can bring about an anomalously large transport coefficient (compared with a random collisional process) of thermodynamic quantities such as number of particles, temperature, pressure or electrical resistivity. In fact, plasma instabilities can be caused by almost any form of thermodynamical nonequilibrium. The "anomalous" transport effects are quite common and the classical transport generated by collision effect is even rarer.

The next important aspect of plasma instabilities exists in the generation of various kinds of waves. Except for special cases (such as that shown in Fig. 1a), plasma instabilities are caused by perturbations that oscillate around some particular range of frequency. In such cases, the instability emits a wave. Because of the nature of the instability, the rate of increase of the wave amplitude is proportional to its instantaneous amplitude; this emission process is much more effective than a single particle emission process such as Cherenkov or synchrotron radiation.

Now we know basically what a plasma instability is and why it is interesting to study. In the rest of this chapter, we will discuss how to find an instability. For this, we will need some physical and mathematical tools to describe the dynamics of plasma behavior. Section 1.2 will thus be devoted to the introduction of basic equations that are useful for the description of plasma dynamics. In Section 1.3 we will introduce the concept of the dispersion relation and its meaning with respect to a plasma instability using the simple example of a two stream instability, and in Section 1.4 we will discuss a classification of various instabilities and general methods of finding plasma instabilities using the dispersion relation.

1.2 Basic Equations

As pointed out in Section 1.1, plasmas in space can be considered effectively collision free (except in the ionosphere or in the solar surface). To confirm this fact it is convenient to introduce the concept of collision frequency. From the elementary text book, we know that the time rate of collision, called collision frequency ν, is given by

$$\nu = v n \sigma \tag{1.1}$$

where v is the speed of the colliding particle, σ is the cross section of the object particle and n is the density of the object particle. In the case of collision between charged particles, the cross section σ can be obtained as follows. For electrons, one might expect to use the classical electron radius $(e^2/(4\pi\varepsilon_0 m_e c^2))$ but because of the Coulomb force, they cannot usually approach so close. Let us see how close two particles with charge e (coulomb) and mass m_1 and m_2 (kg) can approach before being deflected by the Coulomb force. At a distance $r_0\,(m)$ the relative kinetic energy $\bar{m}\bar{v}^2/2$ (total kinetic energy minus that of the center of gravity) becomes comparable to the Coulomb potential energy $e^2/(4\pi\varepsilon_0 r_0)$, where $\bar{m}$ is the effective mass given by $\bar{m} = m_1 m_2/(m_1 + m_2)$, ε_0 is the dielectric constant in free space (8.854×10^{-12} farad/m), $\bar{v}$ is the relative speed of colliding particles, and e is the electron charge (1.602×10^{-19} coulomb). This distance r_0 is the closest distance of approach and is given by

$$r_0 = \frac{e^2}{2\pi\varepsilon_0 \bar{m}\bar{v}^2} \text{ (m)} \tag{1.2}$$

The cross section of the collision σ may therefore be obtained from r_0 as $\sigma = \pi r_0^2$ or

$$\sigma = \frac{e^4}{4\pi\varepsilon_0^2 \bar{m}^2 \bar{v}^4} \text{ (m}^2\text{)}. \tag{1.3}$$

The collision frequency is now obtained by substituting Eq. (1.3) into Eq. (1.1)

$$\nu = \frac{v e^4 n}{4\pi\varepsilon_0^2 \bar{m}^2 \bar{v}^4} \text{ (sec}^{-1}\text{)}. \tag{1.4}$$

For the case of electron-electron collision, Eq. (1.4) can be reduced, aside from a numerical factor of $1/\pi$, to

$$\nu_{ee} \doteq \omega_{pe}/(n \lambda_{De}^3) \text{ (sec}^{-1}\text{)}, \qquad (1.5)\star$$

* Because the Coulomb force extends at least to the Debye length, even a particle passing at a distance much larger than r_0 from an object particle is deflected by some angle. To introduce the correction for this effect, ν_{ee} in Eq. (1.5) must be multiplied by a quantity $\ln \Lambda$, where $\Lambda = \lambda_{De}/r_0 = 4\pi n \lambda_{De}^3$, which is the size of the inverse plasma parameter. Ordinarily, $\ln \Lambda \sim 20$.

where ω_{pe} is the plasma (angular) frequency given by

$$\omega_{pe} = \left(\frac{e^2 n}{\varepsilon_0 m_e}\right)^{1/2} \doteq 56.4 \sqrt{n(\mathrm{m}^{-3})}\ (\mathrm{sec}^{-1}) \tag{1.6}$$

and λ_{De} is called the electron *Debye length* given by

$$\lambda_{De} = v_{Te}/\omega_{pe}\ (\mathrm{m}), \tag{1.7}$$

and v_{Te} and m_e are the electron thermal speed and mass. $n\lambda_{De}^3$ in Eq. (1.5) represents the number of electrons in a box with the size of the Debye length λ_{De}, and λ_{De} represents the distance beyond which the Coulomb force is shielded. For a plasma to behave as a collective body, rather than as an uncorrelated bunch of particles, $n\lambda_{De}^3$ must be much larger than unity; the small quantity $1/(n\lambda_{De}^3)$ is often called the *plasma parameter*.

Therefore one can see in general $\nu_{ee} \ll \omega_{pe}$. Using the expression (1.4), we can also derive a collision frequency between different particles. The electron-ion collision frequency ν_{ei} is comparable to $\nu_{ee}(\nu_{ei} = \nu_{ee}/2\sqrt{2})$ while the ion-ion collision frequency ν_{ii} is given by

$$\begin{aligned}\nu_{ii} &\doteq \omega_{pi}/(n\lambda_{Di}^3) \\ &= \nu_{ee}(m_e T_e^3/m_i T_i^3)^{1/2}\end{aligned} \tag{1.8}$$

where $T_e(=m_e v_{Te}^2)$ and $T_i(=m_i v_{Ti}^2)$ are electron and ion temperatures and m_i and v_{Ti} are the ion mass and thermal speed. The ion-electron collision frequency ν_{ie} is smaller than electron-ion collision frequency by the mass ratio (m_e/m_i) (because ions are difficult to deflect with electrons) and hence is negligible in most cases.

With these preparations, we will now derive the basic equation to represent the dynamics of a plasma with a small plasma parameter $1/(n\lambda_D^3)$. Consider the motion of n number of electrons and ions. Even if there are millions of particles, unlike the people in New York City, they obey only one rule of motion, decided by an equation of motion in which the acceleration is dominated by the electromagnetic force *(Lorentz force)*:

$$m_j \frac{d\boldsymbol{v}_j}{dt} = q_j[\boldsymbol{E}(\boldsymbol{x}_j, t) + \boldsymbol{v}_j \times \boldsymbol{B}(\boldsymbol{x}_j, t)] \tag{1.9}$$

where the subscript j shows j^{th} particle with charge q_j and mass m_j located at $\boldsymbol{x}_j$ at time t. Note that the electric field $\boldsymbol{E}$ and the magnetic field $\boldsymbol{B}$ are evaluated at the position of the particle j but are generated by all the other particles and possibly by some external sources.

Now suppose that the electric and the magnetic field is known as a continuous function of space and time. We can then integrate Eq. (1.9) to obtain $\boldsymbol{x}_j(t)$ (by combining $\dot{\boldsymbol{x}}_j=\boldsymbol{v}_j$) provided we know the initial velocity $\boldsymbol{v}_j(0)$. However, to find the initial velocity of a large number of particles is impossible, therefore it is convenient to introduce a probability distribution function for the initial velocity such that the probability of finding the particle j with an initial velocity range $-\infty<\boldsymbol{v}(0)<\boldsymbol{v}$ at point $\boldsymbol{x}$ is defined by $F_j(\boldsymbol{x}, \boldsymbol{v})$ where $0\leq F_j(\boldsymbol{x}, \boldsymbol{v})\leq 1$. The probability density function $f(\boldsymbol{x}, \boldsymbol{v})$ is obtained by

$$f_j(\boldsymbol{x}, \boldsymbol{v})=\frac{\partial^3 F_j}{\partial v_x \partial v_y \partial v_z}. \tag{1.10}$$

Here $\boldsymbol{x}$ is no longer the position of the particle $\boldsymbol{x}_j$, because we do not know exactly where the jth particle is located. Hence $\boldsymbol{x}$ and $\boldsymbol{v}$ in the above expression are independent variables. Let us now consider how the probability density function f defined in Eq. (1.10) changes in time. It changes in time because of the Lorentz force. This can be found from the equation of conservation of f_j in phase space, $\boldsymbol{x}$, $\boldsymbol{v}$, t.

$$\frac{\partial f_j}{\partial t}+\dot{\boldsymbol{x}}\cdot\frac{\partial f_j}{\partial \boldsymbol{x}}+\dot{\boldsymbol{v}}\cdot\frac{\partial f_j}{\partial \boldsymbol{v}}=0 \tag{1.11}$$

where $\dot{\boldsymbol{x}}=\boldsymbol{v}$, and $\dot{\boldsymbol{x}}\cdot\frac{\partial}{\partial \boldsymbol{x}}=v_x\frac{\partial}{\partial x}+v_y\frac{\partial}{\partial y}+v_z\frac{\partial}{\partial z}=\boldsymbol{v}\cdot\boldsymbol{\nabla}$.

The acceleration $\dot{\boldsymbol{v}}$, is given by the Lorentz force. The Lorentz force acting on the particle consists of two kinds of fields. One is the electromagnetic field generated by the collective motion of all the other particles; the other is that generated by one or possibly more particles that are located within the colliding distance of the particle we are looking at. However, if the plasma parameter $(n\lambda_D^3)^{-1}$ is small, we can show that the effect due to colliding particles is negligible. One way of showing this is to estimate the possibility that a neighboring particle is located within the colliding distance, i.e. r_0 defined in Eq. (1.2). The mean spacing of the particles is $n^{-1/3}$. Thus such a possibility is estimated by the ratio nr_0^3, which becomes, from Eq. (1.2), $1/(n\lambda_D^3)^2$. Therefore if $n\lambda_D^3\gg 1$, it is rather unlikely that a neighboring particle is within the colliding distance. In this case, one can ignore the acceleration due to colliding particles and Eq. (1.11) reduces, with the help of the Lorentz force to

$$\frac{\partial f(\boldsymbol{x}, \boldsymbol{v}, t)}{\partial t}+\boldsymbol{v}\cdot\frac{\partial f(\boldsymbol{x}, \boldsymbol{v}, t)}{\partial \boldsymbol{x}}+\frac{q}{m}[\boldsymbol{E}(\boldsymbol{x}, t)+\boldsymbol{v}\times\boldsymbol{B}(\boldsymbol{x}, t)]\cdot\frac{\partial f}{\partial \boldsymbol{v}}=0 \tag{1.12}$$

where $\boldsymbol{E}$ and $\boldsymbol{B}$ are average electric field intensity and magnetic flux density produced by the collective motion of plasma particles. The above expression is called the *Vlasov equation* and is one of the basic equations we introduce in this section.

Note here that the coordinate variable $\boldsymbol{x}$ is no longer the position of our j^{th} particle but is an *independent variable* fixed to the coordinate space. Although we have derived Eq. (1.12) for a specific particle j, one can now see that there is no need to distinguish the distribution function of any one of the particles in the plasma. Therefore we have deleted the subscript j here. Eq. (1.12) describes the dynamic change of f due to the collective field. The time scale of the change will be of the order ω_p^{-1}, while we ignore the change due to collisions whose time scale is ν^{-1}. The ratio of these two time scales is again $n\lambda_D^3$; in this way one can also justify neglection of the collision effect.

The electromagnetic field in Eq. (1.12) can be obtained from the *Maxwell's equations.*

$\int_{-\infty}^{\infty} f(\boldsymbol{x}, \boldsymbol{v}, t)\,d\boldsymbol{v}$ represents the total probability that a particle can be found at $\boldsymbol{x}$ and t. If we multiply by the space averaged number density n_0, the resultant quantity gives the number density of plasma particles, i.e.,

$$n(\boldsymbol{x}, t) = n_0 \int_{-\infty}^{\infty} f(\boldsymbol{x}, \boldsymbol{v}, t)\,d\boldsymbol{v}. \tag{1.13}$$

Similarly the current density for each species can be obtained by

$$\boldsymbol{J}(\boldsymbol{x}, t) = q n_0 \int_{-\infty}^{\infty} \boldsymbol{v} f(\boldsymbol{x}, \boldsymbol{v}, t)\,d\boldsymbol{v}. \tag{1.14}$$

If we use subscripts e and i to distinguish the distribution functions of electrons and ions, Maxwell's equations (in MKS units) can be written as

$$\nabla \times \boldsymbol{E} = -\frac{\partial \boldsymbol{B}}{\partial t}, \tag{1.15}$$

$$\nabla \times \boldsymbol{B} = \mu_0 e n_0 \left(\int_{-\infty}^{\infty} \boldsymbol{v} f_i\,d\boldsymbol{v} - \int_{-\infty}^{\infty} \boldsymbol{v} f_e\,d\boldsymbol{v} \right) + \frac{1}{c^2} \frac{\partial \boldsymbol{E}}{\partial t}, \tag{1.16}$$

$$\nabla \cdot \boldsymbol{B} = 0, \tag{1.17}$$

$$\nabla \cdot \boldsymbol{E} = \frac{e n_0}{\varepsilon_0} \left(\int_{-\infty}^{\infty} f_i\,d\boldsymbol{v} - \int_{-\infty}^{\infty} f_e\,d\boldsymbol{v} \right), \tag{1.18}$$

where μ_0 is the permeability of free space ($4\pi \times 10^{-7} = 1.257 \times 10^{-6}$ henry/m), c is the speed of light (2.998×10^8 m/sec), and $\int d\boldsymbol{v}$ shows the volume integral in the velocity space.

Using a combination of Maxwell's equations and the Vlasov equation is the most powerful method for describing plasma dynamics. However, as usual, the most powerful method is also most difficult to apply. In particular, the Vlasov equation is difficult to solve in a complex configuration. However, the complex configuration becomes important in many cases only in relation to the macroscopic instabilities generated by non-uniformity in geometric space. In such cases the scale length of concern is usually much larger than the average proton *Larmor radius* $\varrho_i (= v_{Ti}/\omega_{ci})$. Consequently the time scale of the problem can be considered to be much smaller than the ion cyclotron periods, ω_{ci}^{-1}. In these cases we can simplify the Vlasov equation.

The idea is to treat the plasma as if it were a continuous fluid; the resultant equations are called *m*agneto*h*ydro*d*ynamic equations (MHD equations). Exactly speaking such a treatment is only possible when the time scale of the problem is longer than the mean free time, ν^{-1}, so that the plasma behaves as if it were entirely collisional. However, in reality, it may be a good approximation as long as the time scale is longer than the cyclotron period ω_{ci}^{-1}, which is usually much shorter than ν^{-1}. Leaving the justification of this until later, let us briefly show how the equations are derived. In the derivation we assume that the time scale of concern is much longer than the mean free time. In such a low frequency regime, one can assume that the electrons and ions macroscopically move together to form a charge neutral (quasineutral) situation in the plasma. We thus can represent the electron and proton densities by one quantity ⋆

$$\begin{aligned} n(\boldsymbol{x}, t) &\sim n_0 \int_{-\infty}^{\infty} f_e(\boldsymbol{x}, \boldsymbol{v}, t)\, d\boldsymbol{v} \\ &\sim n_0 \int_{-\infty}^{\infty} f_i(\boldsymbol{x}, \boldsymbol{v}, t)\, d\boldsymbol{v} \end{aligned} \tag{1.19}$$

where f_e and f_i represent the electron and proton distribution function.

The first of the MHD equations represents mass conservation, which is obtained simply by integrating the Vlasov equation for ions over velocity space;

$$\frac{\partial n(\boldsymbol{x}, t)}{\partial t} + \nabla \cdot [n(\boldsymbol{x}, t)\, \boldsymbol{v}(\boldsymbol{x}, t)] = 0. \tag{1.20}$$

⋆ There is also a two-fluid model to describe the different dynamics between electrons and ions. We follow, however, the one-fluid model for simplicity.

$\boldsymbol{v}(\boldsymbol{x}, t)$ is the *fluid velocity* and not the particle velocity, and is represented by the average ion velocity given by

$$n(\boldsymbol{x}, t)\, \boldsymbol{v}(\boldsymbol{x}, t) = n_0 \int \boldsymbol{v} f_i(\boldsymbol{v}, \boldsymbol{x}, t)\, d\boldsymbol{v}\,. \tag{1.21}$$

The use of the Vlasov equation even for the present collisional case will be justified later. The next equation is the equation of motion of the fluid. To obtain this we multiply the Vlasov equation by $\boldsymbol{v}$ and integrate over velocity space giving, say, for the x component,

$$\frac{\partial}{\partial t}[m_j n(\boldsymbol{x}, t) v_{jx}(\boldsymbol{x}, t)] + \sum_{\alpha=x,y,z} \frac{\partial}{\partial \alpha}[m_j n(\boldsymbol{x}, t)\langle v_x v_\alpha \rangle_j] - q_j n(\boldsymbol{x}, t)\,[\boldsymbol{E}(\boldsymbol{x}, t) + \boldsymbol{v}_j(\boldsymbol{x}, t) \times \boldsymbol{B}(\boldsymbol{x}, t)]_x = 0 \tag{1.22}$$

where $\langle A \rangle$ shows average defined by

$$n(\boldsymbol{x}, t)\,\langle A \rangle_j = n_0 \int_{-\infty}^{\infty} A f_j\, d\boldsymbol{v} \tag{1.23}$$

and $j = e$ or i.

If we use the mass conservation Eq. (1.20) and the identity,

$$\frac{d}{dt} = \frac{\partial}{\partial t} + (\boldsymbol{v}(\boldsymbol{x}, t) \cdot \nabla) \tag{1.24}$$

and then construct an addition of Eq. (1.22) for ions and electrons, we have, neglecting electron inertia,

$$m_i n(\boldsymbol{x}, t) \frac{d\boldsymbol{v}(\boldsymbol{x}, t)}{dt} = \boldsymbol{J}(\boldsymbol{x}, t) \times \boldsymbol{B}(\boldsymbol{x}, t) - \nabla p(\boldsymbol{x}, t) \tag{1.25}$$

where $\boldsymbol{J}$ is the total current density, $\boldsymbol{J}_i + \boldsymbol{J}_e$, and p is the pressure defined by

$$\begin{aligned} p &= n\,[m_i \langle (v_x - \langle v_x \rangle)^2 \rangle_i + m_e \langle (v_x - \langle v_x \rangle)^2 \rangle_e] \\ &= \frac{n}{3}\,[m_i \langle (\boldsymbol{v} - \langle \boldsymbol{v} \rangle)^2 \rangle_i + m_e \langle (\boldsymbol{v} - \langle \boldsymbol{v} \rangle)^2 \rangle_e] \\ &= n(T_i + T_e) \quad \text{(for Maxwell distribution; } T \text{ in energy units).} \end{aligned} \tag{1.26}$$

In the derivation of Eq. (1.25), we assumed an isotropic pressure, which is justified if several collisions occur within the time scale of concern, because collision tends to maintain the isotropic velocity distribution function and hence the isotropic pressure.

If we construct the difference of Eq. (1.22) and assume a small difference in electron and ion temperatures, we have an equation called *Ohm's law*, i.e.,

$$\boldsymbol{E}(\boldsymbol{x}, t)+\boldsymbol{v}(\boldsymbol{x}, t)\times\boldsymbol{B}(\boldsymbol{x}, t)=\eta\boldsymbol{J}(\boldsymbol{x}, t) \tag{1.27}$$

where η is the electrical resistivity given by

$$\eta=\frac{m_e\nu_{ei}}{e^2 n_0}=\frac{\nu_{ei}}{\varepsilon_0\,\omega_{pe}^2}\ (\text{ohm}\cdot\text{meter}). \tag{1.28}$$

In derivation of Eq. (1.27), the inertia term $\partial/\partial t$ was ignored because we were considering only low frequency phenomena with $\omega\ll\omega_{ci}$. The reader may now be puzzled as to why the electron collision frequency appeared in the right hand side of Eq. (1.27). In fact if we use the Vlasov equation this term ($\eta\boldsymbol{J}$) does not appear. We have introduced this term to be consistent with our basic assumption of $\omega\ll\nu$. On the other hand, even if $\omega\ll\nu$ the mass conservation equation is valid because the Coulomb collision does not destroy the mass. In a similar way, the fluid equation of motion (1.25) is justified because the total momentum is conserved by the collision. Because of these reasons, we have used the Vlasov equation (which ignores collisions) to derive Eqs. (1.20) and (1.25). However, since the current originates because of the difference between electron and ion fluid velocities, the collision effect does not cancel out in the averaging process, and the resistivity in Eq. (1.27) must in general be retained.

The final equation we need is the equation of state which controls the dynamic change of the plasma pressure p. If many collisions occur during one cycle of plasma motion, the velocity distribution function essentially retains its shape. Because of this we assume the distribution function to be symmetric around its average (not necessarily a Maxwell distribution). With this assumption, the pressure change will be adiabatic; $p\sim n^\gamma$. The value of γ is obtained by multiplying the Vlasov equation by v^2, averaging over velocity space and combining Eqs. (1.20), (1.25) and the assumption of the symmetric distribution. The value of γ thus obtained is 5/3. Thus we have obtained the total set of MHD equations, which we repeat here

$$\frac{\partial n}{\partial t}+\nabla\cdot(n\boldsymbol{v})=0, \tag{1.20}$$

$$m_i n \frac{d\boldsymbol{v}}{dt} = \boldsymbol{J} \times \boldsymbol{B} - \nabla p, \tag{1.25}$$

$$\boldsymbol{E} + \boldsymbol{v} \times \boldsymbol{B} = \eta \boldsymbol{J}, \tag{1.27}$$

$$\frac{dp}{dn} = \frac{5}{3} \frac{p}{n}. \tag{1.29}$$

The matching Maxwell's equations are, taking into account the quasi-neutrality situation,

$$\nabla \times \boldsymbol{E} = -\frac{\partial \boldsymbol{B}}{\partial t}, \tag{1.15}$$

$$\nabla \times \boldsymbol{B} = \mu_0 \boldsymbol{J}, \tag{1.30}$$

$$\nabla \cdot \boldsymbol{B} = 0. \tag{1.17}$$

Eq. (1.18) is not needed. Although these MHD equations are obtained for a frequency range much lower than the collision rate, they are useful even at a higher frequency, say, up to the proton cyclotron frequency, if the symmetric distribution function is justified during the process of a problem. In many cases, the possible existence of an anomalous collision rate generated by a microscopic instability or superthermal fluctuations, which can become comparable to ω_{ci}, can justify such a distribution function. Under these circumstances, however, one must be careful in evaluating the effective resistivity η and the net loss of fluid momentum to the wave, which will modify Eqs. (1.25) and (1.27).

The use of MHD equations may be justified even though the plasma is locally stable, if there exist fluctuating fields whose intensity is much larger than the thermodynamical equilibrium level.* Diffusion of particle orbits due to these fluctuating fields could maintain the symmetry and isotropy of the distribution function.

When the magnetic field intensity is large (more precisely when $\omega_{ce} \gg \nu_{ei}$)**, $\eta \boldsymbol{J}$ in Eq. (1.27) can be neglected. Such a circumstance is called the "Frozen in" state, because the plasma and the magnetic field can be shown to move together in such a case. It is important to realize that a plasma instability, by enhancing the resistivity, can destroy this "Frozen in" condition.

---

* At thermodynamic equilibrium, the fluctuating field energy density $\varepsilon_0 E^2/2$ is approximately given by $nT/(n\lambda_D^3)$, where T is the plasma temperature in energy units.

** If we use $-i\omega m_i n \boldsymbol{v} \sim \boldsymbol{J} \times \boldsymbol{B}$, the ratio $|\eta \boldsymbol{J} / \boldsymbol{v} \times \boldsymbol{B}|$ becomes $\sim \omega \nu_{ei} / \omega_{ci} \omega_{ce}$.

The two sets of equations, Vlasov and MHD, are applicable to most of the plasmas in space. Readers are encouraged, however, to check their applicability carefully for each specific problem.

In the ionosphere, where collisions with neutrals are more frequent, neither of these equations is applicable. Here we use isothermal diffusion-type equations of motion. We leave this case to the relevant chapter.

1.3 Dispersion Relations

In this section, we present a general concept of plasma instabilities. The definition of instability will be discussed as well as ways of finding and analyzing instabilities.

As in most cases, the general concept is more easily understood by considering a typical example. We consider here two-stream instability in a system of an electron stream with velocity $\boldsymbol{v}_0$ and a stationary plasma. If we restrict ourselves to a frequency range much higher than the ion plasma frequency, we can ignore ion dynamics. (We shall discuss the case where ion dynamics are involved in Section 2.2).

In the example we are considering here, if the electron temperature is high so that the thermal velocity v_T is compararable with or larger than v_0 or the phase velocity of the wave v_p, we have to use the Vlasov equation. However, if we consider a cold electron stream in a cold plasma ($v_T=0$), a fluid equation of motion is sufficient. Here the electron stream and the cold plasma form the two streams.

We first consider the dynamics of the stream electrons. The equation of motion of the *electron fluid* is

$$m_e n(d\boldsymbol{v}/dt) = -en(\boldsymbol{E}+\boldsymbol{v}\times\boldsymbol{B}). \tag{1.31}$$

Because we do not know beforehand the kind of electromagnetic field produced by the stream electrons, we have to include arbitrary fields in the equation of motion.

Because the electron stream is moving with respect to the stationary frame of the plasma, it is convenient to write the total derivative in Eq. (1.31) in partial derivatives in time and space, i.e.

$$\frac{d\boldsymbol{v}}{dt} = \frac{\partial \boldsymbol{v}}{\partial t} + (\boldsymbol{v}\cdot\nabla)\boldsymbol{v} \tag{1.32}$$

where

$$\boldsymbol{v}\cdot\nabla = v_x\frac{\partial}{\partial x} + v_y\frac{\partial}{\partial y} + v_z\frac{\partial}{\partial z}. \tag{1.33}$$

This representation is needed because Eq. (1.31) represents the fluid motion and not a single particle motion. When one is concerned with a linear instability, the standard procedure at this point is *linearization*. We write the dependent quantities in terms of *dc* or slowly varying quantities (subscript 0) and *ac* or rapidly varying quantities (subscript 1). For example, for the velocity of the electron stream $\boldsymbol{v}$

$$\boldsymbol{v} = \boldsymbol{v}_0 + \boldsymbol{v}_1 . \tag{1.34}$$

We consider the quantity with subscript 1 to be a perturbation to the state represented by subscript 0. Hence $|\boldsymbol{v}_0| \gg |\boldsymbol{v}_1|$.

Eq. (1.31) for order zero is then

$$\frac{\partial \boldsymbol{v}_0}{\partial t} + (\boldsymbol{v}_0 \cdot \nabla)\boldsymbol{v}_0 = -\frac{e}{m_e}(\boldsymbol{E}_0 + \boldsymbol{v}_0 \times \boldsymbol{B}_0). \tag{1.35a}$$

If the stream is stationary in time and spatially uniform, $\partial \boldsymbol{v}_0/\partial t$ as well as $\partial \boldsymbol{v}_0/\partial \boldsymbol{x}$ vanish. This means that we are considering an infinitely extended stream of electrons. Such a scheme is allowed if the beam current is weak and the cross section of the beam is much larger than the unstable wavelength which will be found later. Then it is immediately obvious that $\boldsymbol{v}_0$ can have an arbitrary magnitude parallel to $\boldsymbol{B}_0$, or is $\boldsymbol{E}_0/\boldsymbol{B}_0$ if it is perpendicular to $\boldsymbol{B}_0$. The perpendicular velocity produced by a *dc* electric field $\boldsymbol{E}_0$ should be the same for both the stream and the stationary electrons; hence in the frame of $\boldsymbol{E}_0/\boldsymbol{B}_0$ there will be no streaming perpendicular to $\boldsymbol{B}_0$. (The $\boldsymbol{E}_0 \times \boldsymbol{B}_0$ drift however does produce two streaming between electrons and ions if the collisions with neutrals exist as will be discussed in Section 2.5.)

Thus in the collisionless case, two streaming is possible only in the direction *parallel to* $\boldsymbol{B}_0$. We set

$$\boldsymbol{v}_0 = v_0 \boldsymbol{e}_{\|} \tag{1.35b}$$

where $\boldsymbol{e}_{\|}$ is the unit vector parallel to $\boldsymbol{B}_0$. The zeroth-order solution is called that of the equilibrium state. When one considers an instability, it is very important to find the equilibrium state first. It is meaningless to discuss the instability of a state where no equilibrium state exists (exceptions are allowed for some limited cases, where the initial non-equilibrium state becomes unstable with a growth rate much faster than the rate of approach to equilibrium).

We now employ a rectangular coordinate system and take the z axis in the direction of the *dc* magnetic field $\boldsymbol{B}_0$. Then the first-order equation for equation of motion (1.31) becomes

$$\frac{\partial \boldsymbol{v}_1}{\partial t} + v_0 \frac{\partial \boldsymbol{v}_1}{\partial z} = -\frac{e}{m_e}(\boldsymbol{E}_1 + \boldsymbol{v}_0 \times \boldsymbol{B}_1 + \boldsymbol{v}_1 \times \boldsymbol{B}_0). \tag{1.36}$$

The next step usually taken after linearization is Fourier-Laplace transformation of the dependent variables, for example for $\boldsymbol{v}_1(\boldsymbol{x}, t)$

$$\boldsymbol{v}_1'(\omega, \boldsymbol{k}) = \int_0^\infty dt \int_{-\infty}^{\infty} d\boldsymbol{x}\, \boldsymbol{v}_1(\boldsymbol{x}, t)\, e^{i(\omega t - \boldsymbol{k}\cdot\boldsymbol{x})} \tag{1.37}$$

where $\boldsymbol{v}_1(\boldsymbol{x}, t)$ is given by

$$\boldsymbol{v}_1(\boldsymbol{x}, t) = \frac{1}{(2\pi)^4} \int_{-\infty + i\sigma}^{\infty + i\sigma} d\omega \int_{-\infty}^{\infty} d\boldsymbol{k}\, \boldsymbol{v}_1'(\omega, \boldsymbol{k})\, e^{i(\boldsymbol{k}\cdot\boldsymbol{x} - \omega t)}. \tag{1.38}$$

For the stability analysis, we usually look for an instability not depending on any initial condition of the perturbations. If we set all the initial values to zero, the transformed results have identically the same form as the one obtainable by substituting a complex amplitude function defined, for example, by

$$\boldsymbol{v}_1(\boldsymbol{x}, t) = \boldsymbol{v}_1''\, e^{i(\boldsymbol{k}\cdot\boldsymbol{x} - \omega t)}. \tag{1.39}$$

Note however, that, $\boldsymbol{v}_1'$ in Eq. (1.37) and $\boldsymbol{v}_1''$ in Eq. (1.39) have different dimensions. It is also important to remember that the transformation in time in Eq. (1.37) is valid for the plane $\mathrm{Im}\,\omega > 0$, so that the integration over time can converge at $t \to +\infty$. This fact will be used in Chapter 2. With these two points in mind, one can obtain the transformed result equivalently by substituting the amplitude function defined in Eq. (1.39). The result then becomes

$$i(k_z v_0 - \omega)\boldsymbol{v}_1 = -\frac{e}{m_e}(\boldsymbol{E}_1 + \boldsymbol{v}_0 \times \boldsymbol{B}_1 + \boldsymbol{v}_1 \times \boldsymbol{B}_0) \tag{1.40a}$$

(for simplicity, we delete prime or double prime for the transformed quantity).

At this stage it is usually convenient to consider the direction of wave propagation. This is determined by the direction of the wave vector $\boldsymbol{k}$ used in the transformation. Here we take $\boldsymbol{k}$ to be in the z direction; then $\boldsymbol{v}_0$, $\boldsymbol{B}_0$, and $\boldsymbol{k}$ are parallel to each other. By doing this we lose an electrostatic cyclotron wave instability (for example, Briggs, 1964) and an electromagnetic instability (Weibel, 1959) but gain a considerable simplification, which is important for presenting an example. The equation of continuity shows that $\boldsymbol{v}_1$ is directed also in the z direction, whereas Poisson's equation gives $\boldsymbol{E}_1$ in the z direction. This means there exists no perturbed magnetic field $\boldsymbol{B}_1$ in this case. $\boldsymbol{v}_1 \times \boldsymbol{B}_0$ also vanishes because $\boldsymbol{v}_1$ is parallel to $\boldsymbol{B}_0$. Eq. (1.40a) then simplifies to

$$v_1 = \frac{-E_1\, e/m_e}{i(k v_0 - \omega)} \tag{1.40b}$$

where v_1, E_1, and k are in the z direction.

Now we have to obtain E_1. The sources of the electromagnetic fields are current density and charge density. We have to derive them from the velocity in Eq. (1.40b). One equation needed for this is the relation of the current density J to the particle velocity v and number density n: $J = -env$, which, in the linearized form, becomes (for an electron stream)

$$J_1 = -e(n_0 v_1 + n_1 v_0). \tag{1.41}$$

We also need the equation of continuity

$$-e\frac{\partial n_1}{\partial t} + \frac{\partial J_1}{\partial z} = 0 \tag{1.42a}$$

or

$$e\omega n_1 + k J_1 = 0. \tag{1.42b}$$

By eliminating v_1 and n_1 from Eqs. (1.40b), (1.41), and (1.42b), we can express J_1 in terms of E_1 and dc quantities as

$$J_1 = -i\omega\varepsilon_0 \left[\frac{-\omega_{ps}^2}{(\omega - k v_0)^2}\right] E_1 \tag{1.43}$$

where ω_{ps} is the plasma frequency of the stream electrons.

Now we introduce the concept of an equivalent dielectric constant. In the second curl equation of Maxwell's equations

$$\nabla \times \boldsymbol{H} = \boldsymbol{J} + \varepsilon_0 \frac{\partial \boldsymbol{E}}{\partial t}. \tag{1.44}$$

If we know the relation between the current density $\boldsymbol{J}$ and electric field intensity $\boldsymbol{E}$, such as shown in Eq. (1.43), the right-hand side of (1.44) may be equivalently written as

$$\nabla \times \boldsymbol{H} = -i\omega\varepsilon_0(1+\varepsilon)\boldsymbol{E}. \tag{1.45a}$$

ε expressed as in Eq. (1.45a), which is a tensor in general, is called the *equivalent dielectric constant (tensor)*.★ In the case of the electron stream, ε is given from Eq. (1.43) by

$$\varepsilon = -\frac{\omega_{ps}^2}{(\omega - k v_0)^2} (\equiv \varepsilon_s). \tag{1.46}$$

Exactly in the same way, one can derive the equivalent dielectric constant, ε_p, for the stationary plasma electrons

$$\varepsilon = -(\omega_p^2/\omega^2) \equiv \varepsilon_p \tag{1.47}$$

where ω_p is the plasma frequency of the stationary electrons. Because there exists no perturbed magnetic field, Eq. (1.45a) reduces to

$$J_1 - i\omega\varepsilon_0 E_1 = 0. \tag{1.45b}$$

If we substitute the sum of the perturbed currents in the stream and in the plasma for J_1 in Eq. (1.45b), we have

$$-i\omega\varepsilon_0(1+\varepsilon_s+\varepsilon_p)E_1 = 0. \tag{1.45c}$$

The nontrivial solution of Eq. (1.45c) is given by

$$D(\omega, k) \equiv \omega(1+\varepsilon_s+\varepsilon_p) = 0. \tag{1.48a}$$

Eq. (1.48a) is called the dispersion relation.

Let us now briefly review the process we have used to derive the dispersion relation. We first assumed an arbitrary electromagnetic field and obtained the response in particle motion produced by the Lorentz force. We linearized the equation by assuming that the perturbation is infinitely small. We then obtained the electromagnetic field produced by the charge and the current distribution that appear as a consequence of the initially assumed electromagnetic field. The dispersion relation represents a relation between ω and $\boldsymbol{k}$ that makes the assumed electro-

★ In many cases, the equivalent dielectric constant is defined by $1+\varepsilon$ rather than by ε itself, that is by including the free space dielectric constant (cf. Chapter 4). In these cases ε is called the susceptibility. We use the definition without the free space dielectrics, here.

magnetic field consistent with the induced field for an infinitely small perturbation of a form $e^{i(\boldsymbol{k}\cdot\boldsymbol{x}-\omega t)}$. Thus if the dispersion relation gives a root for ω with positive imaginary part, the self-consistent field grows exponentially in time.

1.4 Analysis of Instability

A plasma is called linearly unstable when the dispersion relation

$$D(\omega, \boldsymbol{k})=0 \tag{1.49}$$

has a solution for ω with a positive imaginary part for any real value of $\boldsymbol{k}$.

We will discuss the nature of the dispersion relation given in Eq. (1.49) in the context of a stability analysis (Hasegawa, 1968). In the particular example of a stream of electrons and a stationary plasma, the dispersion relation has the following form (from Eqs. (1.46) and (1.47))

$$1-\frac{\omega_p^2}{\omega^2}-\frac{\omega_{ps}^2}{(\omega-k v_0)^2}=0. \tag{1.48b}$$

Although Eq. (1.48b) is a fourth-order algebraic equation for ω, it can easily be seen that two of the roots can become complex by plotting the left-hand side versus ω and counting the number of zero crossings. It is also easy to see that if $v_0=0$, the complex roots disappear; thus the instability arises from a relative velocity between two groups of electrons.

In this section, however, we are not interested only in the fact that a two-stream flow of electrons becomes unstable. We are interested in a more general theory of the nature of the dispersion relation that leads to an instability.

The fact that the instability is generated by the ε_s part of the dispersion relation implies that ε_s has a particular nature as a dielectric constant. According to Landau and Lifshitz (1960), the electric field energy density W of a wave propagating in a weakly dissipative dielectric medium can in general be expressed as

$$W=\frac{\partial[\omega\varepsilon_0(1+\varepsilon)]}{\partial\omega}\frac{\langle E^2\rangle}{2}=\omega\varepsilon_0\frac{\partial\varepsilon}{\partial\omega}\frac{\langle E^2\rangle}{2} \tag{1.50}$$

where $\langle E^2\rangle$ is the time average of the square of the electric-field amplitude. If we use this relation, the energy of a wave in a stream W_s, and in a plasma W_p, can be calculated, respectively as

$$W_s \propto \frac{\partial\,\omega(1+\varepsilon_s)}{\partial\,\omega} = \omega\,\frac{\partial\,\varepsilon_s}{\partial\,\omega} = \frac{2\,\omega\,\omega_{ps}^2}{(\omega - k\,v_0)^3}\,,$$

$$W_p \propto \frac{\partial\,\omega(1+\varepsilon_p)}{\partial\,\omega} = \omega\,\frac{\partial\,\varepsilon_p}{\partial\,\omega} = \frac{2\,\omega_p^2}{\omega^2}\,, \tag{1.51}$$

where use is made of the fact that the dispersion relation $1+\varepsilon=0$ is satisfied for each species.

Thus we can see that W_p is always positive, whereas W_s can be *negative* if $\omega < k\,v_0$. Now the dispersion relation for stream electrons alone can be obtained from Eq. (1.48b) by setting $\omega_p = 0$ as

$$1+\varepsilon_s = 0$$

or

$$\omega - k\,v_0 = \pm\,\omega_{ps}\,. \tag{1.51}$$

Therefore, the mode corresponding to the lower sign of Eq. (1.51) (called the slower wave) does in fact satisfy the relation $\omega < k\,v_0$. Hence this mode carries a *negative energy,* and the corresponding wave is called the *negative-energy wave* (Chu, 1951). Note that this is the consequence of the fact that energy is not Galilean invariant. The present two-stream instability can be interpreted as being caused by the coupling between the negative-energy wave in the stream and the positive-energy wave in the plasma.

To see this point more clearly, let us plot the dispersion relation in the $\omega - k$ plane. We plot the dispersion relation of the electron stream given by Eq. (1.51) and that of the stationary plasma given by $\omega = \omega_p$ *separately* (but in the same plane) as shown in Fig. 2. We can see that the plasma wave dispersion line crosses those of the faster (upper sign in Eq. (1.51)) and slower waves of the stream at points A and B. At each of these points the pair of the crossing waves have an identical phase velocity (ω/k) and wave frequency; hence, we can expect a coupling of these pairs. As a consequence of the coupling, the dispersion relation of each of these waves is modified. If we plot the total dispersion relation shown in Eq. (1.48b), we can see how such a modification takes place. In general, a set of two dispersion lines never remain crossed when the coupling takes place unless these lines represent orthogonally polarized waves. In the present example, the modification takes place as shown by broken lines in this figure. Interestingly enough, the nature of the modification is different at points A and B; at point A, it splits up parallel to ω axis, while at point B, parallel to k axis. The readers are encouraged to check this.

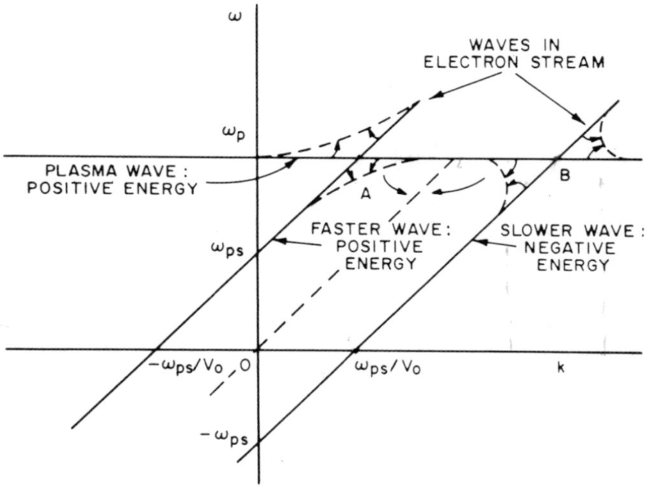

Fig. 2. Dispersion relations of coupling between an electron plasma oscillation and slower and faster waves on electron stream

At $\omega < \omega_p$, if we solve for a complex k the two lines representing the slower and the faster waves in the stream merge into straight line $\omega = k v_0$ shown by the broken line. The solution of the dispersion relation gives a complex ω for a real k or a complex k for a real ω in this regime.

If we look at the resultant dispersion relation, we can see that at point A, we still have real ω solutions for a given value of wave number k near this point, while at point B, ω becomes complex for the value of k in this vicinity. This shows that, by coupling to the plasma oscillation, the negative energy wave on the stream becomes unstable.

If the coupling is weak enough so that it occurs between only two modes (unfortunately the above example is not this case, because three waves whose dispersion relations are given by $\omega = k v_0 \pm \omega_{ps}$, and $\omega = \omega_p$ are coupling simultaneously), four different cases can occur as shown in Fig. 3 (Sturrock, 1958). Plotted in this figure in $\omega - k$ coordinates are portions of the dispersion relation that represent couplings between two modes of different characteristics.

Cases *a* and *b* are for two waves with opposite signs of energy, whereas *c* and *d* are for two waves with the same sign in energy. In case *a* or *b*, a complex ω solution results for real k, hence representing an unstable situation; in case *c* or *d*, ω is always real, thus representing a stable situation. The difference between *a* and *b* (as well as *c* and *d*) is in the sign of the group velocity ($v_g = \partial\omega/\partial k$) of the two coupling modes.

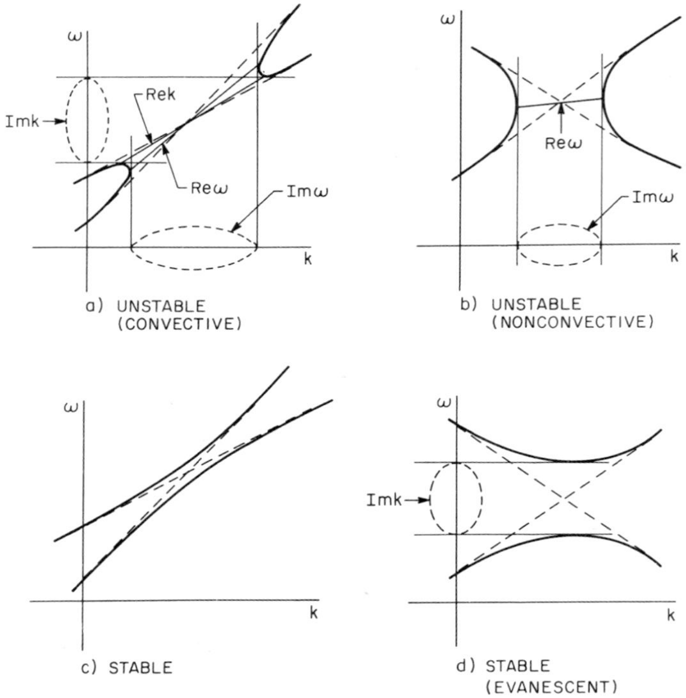

Fig. 3a–d. Dispersion diagrams for two coupled waves of various natures: (a) opposite energy, same group velocity; (b) opposite energy, opposite group velocity; (c) same energy, same group velocity; and (d) same energy, opposite group velocity

Specifically, cases *a* and *c* are for the same sign in the group velocity, but *b* and *d* are for opposite signs.

In case *a*, *k* also can become complex; thus a spatial amplification of a wave at fixed real frequency ω is possible. Such a case is called *convective instability*. In case *b*, the wave number always remains real, hence the instability does not convect away. Such a case is called *non-convective* or *absolute instability*.

It can be shown in general (for example, Hasegawa, 1968) that in a lossless system the necessary condition for an instability to occur (for the root of the dispersion relation to have $\operatorname{Im}\omega > 0$) is that a part of the dielectric constant of a plasma (for example, in the previous case, ε_s of Eq. (1.48a)), should satisfy the condition of a negative-energy wave, i.e.

$$\frac{\partial \omega \varepsilon_0 (1+\varepsilon)}{\partial \omega} < 0 \tag{1.52}$$

for real ω.

We now consider a situation in which losses in the system, either due to collisions or wave-particle interactions, are not negligible. In such a case the necessary condition for an instability is that a part of the dielectric constant satisfies (Hasegawa, 1968)

$$\mathrm{Re}\,\sigma \equiv \mathrm{Re}(-i\omega\varepsilon_0\varepsilon) \propto \mathrm{Im}\,\varepsilon < 0 \tag{1.53}$$

for real ω, where $\sigma(=-i\omega\varepsilon_0\varepsilon)$ represents the equivalent conductivity of a plasma. Eq. (1.53) represents a condition of negative real conductivity, or, in other words, negative dissipation. One can produce Eq. (1.52) from Eq. (1.53) for a case with a small loss by expanding σ in powers of small $\mathrm{Im}\,\omega(>0)$ and requiring $\mathrm{Re}\,\sigma<0$. Thus Eq. (1.53) represents a more general necessary condition for an instability than Eq. (1.52).

We will now see how such cases arise by using again the example of two-stream electrons in two extreme cases. In the first case, the stream electrons are collision-dominated, whereas in the second case the stationary electrons are collision-dominated. Both cases produce instability, but the difference in physical mechanism is worth appreciating.

The collision effect for a cold plasma can be brought in simply by introducing a Langevin-type friction term, νv, into the equation of motion, Eq. (1.31), where ν is an equivalent momentum-transfer collision frequency between electrons and other species. In the first case, where the stream electrons are collision-dominated, ε_s is modified to

$$\varepsilon_s = \frac{i\omega_{ps}^2}{\nu(\omega - k v_0)}, \tag{1.54}$$

where the recombination rate is assumed to be zero. From Eq. (1.54) we can immediately see that when $\omega - k v_0 < 0$, ε_s satisfies the condition (1.53), the condition of negative dissipation. According to the theory presented, the system can become unstable in spite of the collisional dissipation. To see this we write down the entire dispersion relation (assuming no collisions for plasma electrons), following Eq. (1.48a)

$$1 - \frac{\omega_p^2}{\omega^2} + \frac{i\omega_{ps}^2}{\nu(\omega - k v_0)} = 0. \tag{1.55}$$

If we assume for simplicity $\omega_{ps} \ll \omega_p$, the solution of Eq. (1.55) can be written as

$$\omega = \omega_p - i\,\frac{\omega_p\,\omega_{ps}^2}{2\nu(\omega_p - k v_0)}. \tag{1.56}$$

Hence for $k > \omega_p / v_0$, the instability results.

Now we consider the second case where the plasma electrons are collision-dominated. In this case, the dielectric constant for the plasma electrons becomes

$$\varepsilon_p = i\omega_p^2/\nu\omega . \tag{1.57}$$

Naturally $\operatorname{Im}\varepsilon_p > 0$, and the dielectric constant represents a dissipative medium. The full dispersion relation now reads

$$1 - \frac{\omega_{ps}^2}{(\omega - k v_0)^2} + \frac{i\omega_p^2}{\nu\omega} = 0 \tag{1.58}$$

and the approximate solution is obtained for $k v_0 \sim \omega_p$,

$$\omega = k v_0 \pm \omega_{ps}\left(1 - \frac{i\omega_p}{2\nu}\right). \tag{1.59}$$

The lower sign (corresponding to the negative-energy wave) has a solution with $\operatorname{Im}\omega > 0$. Eq. (1.59) implies that a *negative energy wave can cause instability by coupling not only to a positive-energy wave but also to a dissipative medium* (Birdsall *et al.*, 1953). This is because the amplitude of the negative energy wave grows by dissipating its energy.

With these preparations, we can now derive an important general theory. If we define an equivalent longitudinal (electrostatic) plasma conductivity by

$$\sigma(\omega, k) = \mathrm{i}\omega q n_1/(k^2 \varphi_1), \tag{1.60}$$

where q is the relevant particle charge and φ_1 is the perturbed electrostatic potential defined by

$$-i\boldsymbol{k}\varphi_1 = \boldsymbol{E}_1 , \tag{1.61}$$

and $\boldsymbol{k}$ is the vector wave number, the dispersion relation of an electrostatic mode can in general be written as (Hasegawa, 1968)

$$-i\omega\varepsilon_0 + \sigma(\omega, k) = 0 \tag{1.62}$$

where $k = |\boldsymbol{k}|$.

We consider a situation where the system has only a small loss. Mathematically this means that for a real frequency ω and wave number k, $|\operatorname{Re}\sigma| \ll |\omega_r(\partial \operatorname{Im}\sigma/\partial\omega)|$. We also assume that Eq. (1.62) is satisfied by the real frequency $\omega(=\omega_r)$ at real wave number $k(=k_r)$ such that*

$$-i\omega_r\varepsilon_0+i\,\mathrm{Im}[\sigma(\omega_r,k_r)]=0. \tag{1.63}$$

The imaginary part of $\omega(=\omega_i)$ can be obtained in the following way. We expand the dispersion relation Eq. (1.62) around $\omega=\omega_r$ in powers of $i\omega_i$

$$0=-i\varepsilon_0(\omega_r+i\omega_i)+i\,\mathrm{Im}[\sigma(\omega_r,k_r)]+\mathrm{Re}[\sigma(\omega_r,k_r)] \\ +i\omega_i\frac{i\partial\,\mathrm{Im}\,[\sigma(\omega_r,k_r)]}{\partial\omega_r}. \tag{1.64}$$

$\omega_i(>0)$ is called the growth rate of an instability. Using Eq. (1.63) we have from Eq. (1.64)

$$\omega_i\equiv\mathrm{Im}\,\omega=\frac{\mathrm{Re}\,[\sigma(\omega_r,k_r)]}{\partial\,\mathrm{Im}\,[\sigma(\omega_r,k_r)]/\partial\omega_r-\varepsilon_0}. \tag{1.65a}$$

Eq. (1.65a) is quite a useful relation though it applies only in limited situations with small loss. We have used σ and ε alternatively because σ is more natural when we talk about loss, whereas ε is more natural when we talk about energy. The relation between σ and ε is

$$-i\omega\varepsilon_0\varepsilon=\sigma, \tag{1.66a}$$

hence

$$\mathrm{Re}\,\sigma=\varepsilon_0\,\mathrm{Im}\,(\omega\varepsilon), \\ \mathrm{Im}\,\sigma=-\varepsilon_0\,\mathrm{Re}\,(\omega\varepsilon). \tag{1.66b}$$

Therefore, using the dispersion relation (1.62),

$$\partial\,\mathrm{Im}\,\sigma/\partial\omega-\varepsilon_0=-\mathrm{Re}(1+\varepsilon)\varepsilon_0-\varepsilon_0\omega\partial\,\mathrm{Re}\,\varepsilon/\partial\omega=-\varepsilon_0\omega\partial\,\mathrm{Re}\,\varepsilon/\partial\omega,$$

(1.65a) can alternatively be written as

$$\omega_i=-\frac{\mathrm{Im}\,\varepsilon}{\partial\,\mathrm{Re}\,\varepsilon/\partial\omega}. \tag{1.65b}$$

From Eq. (1.65b), one can conclude immediately that instability results either when a negative dissipation ($\mathrm{Re}\,\sigma<0$) couples to a positive-energy

* As seen in the example of the two stream instability, there are many cases in which Eq. (1.63) itself has complex ω roots. We avoid such cases here.

wave ($\partial \varepsilon/\partial \omega > 0$), such as the example in case 1, or when a negative energy wave ($\partial \varepsilon/\partial \omega < 0$) couples to a positive dissipation (Re $\sigma > 0$), as in case 2.

One can generalize the argument to a general dispersion relation (without restriction to an electrostatic mode) given by

$$D(\omega, \boldsymbol{k}) = 0 \tag{1.67}$$

if the following condition is satisfied by D: that there exists a real $\omega (= \omega_r)$ for a real $\boldsymbol{k}$ such that

$$\begin{aligned} &1. \qquad \operatorname{Re}[D(\omega_r, \boldsymbol{k})] = 0, \\ &2. \qquad |\operatorname{Im}[D(\omega_r, \boldsymbol{k})]| \ll \left| \omega_r \frac{\partial \operatorname{Re}[D(\omega_r, \boldsymbol{k})]}{\partial \omega_r} \right| . \end{aligned} \tag{1.68}$$

Using the same technique as before, by expanding D around $\omega = \omega_r$, we have

$$\omega_i = -\frac{\operatorname{Im}[D(\omega_r, \boldsymbol{k})]}{\partial \operatorname{Re}[D(\omega_r, \boldsymbol{k})]/\partial \omega} . \tag{1.69}$$

In the magnetosphere, because there are different groups of plasmas (in terms of average energy) a situation often arises when a wave propagated by one group (typically the cold-or low-energy group) resonates with particles in other groups (low-energy or medium-to high-energy group) and energy is exchanged between groups. Under these circumstances, the condition of wave propagation (Eq. (1.63) or the first equation of (1.68)) is satisfied by the former group, whereas Reσ in Eq. (1.65a) or ImD in Eq. (1.69) is decided by the latter group. Such an approach simplifies the analysis significantly. Therefore, expressions (1.65) or (1.69) are quite useful in magnetospheric plasmas.

Given the dispersion relation in the form of Eq. (1.67), finding a linear plasma instability requires only the algebra of finding a root for ω with a positive imaginary part for a given value of real wave number $\boldsymbol{k}$. The above argument applies only for a case of a small growth rate. For a case with larger growth rate, a classic technique called Nyquist's theorem is useful. Consider the following Cauchy integral,

$$I = \int_{\omega} \frac{d\omega}{D(\omega, \boldsymbol{k})} \times \frac{dD(\omega, \boldsymbol{k})}{d\omega} \tag{1.70}$$

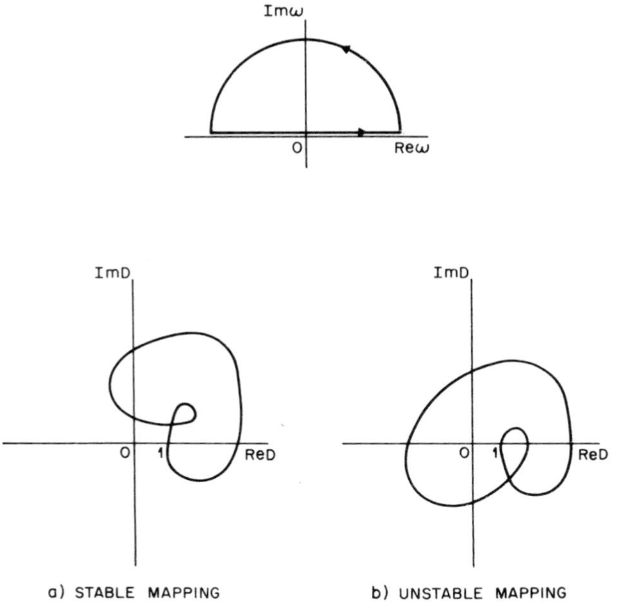

Fig. 4a and b. Nyquist diagram for (a) stable and (b) unstable cases

where the integration contour is along the border of the upper half-plane in the complex ω plane, namely, from $-\infty$ to $+\infty$ on the real ω axis and on an infinite semicircle in the upper half-plane (Fig. 4). If $D=0$ occurs for ω with a positive imaginary part, I has a finite value according to the Cauchy residue theorem. (D is assumed to have no pole. The assumption is shown to be valid for a physical system using causality arguments.) Now, the integral I can be transformed into a D plane integral such as

$$I=\int_{\omega}\frac{dD/d\omega}{D}\,d\omega=\int_{D}\frac{dD}{D}\,, \tag{1.71}$$

where the integration contour is mapped into the D plane. In the D plane, the pole occurs at $D=0$, hence I has a value when the mapped contour in D plane encircles the origin in D plane. In other words, an instability results when the mapping into the D plane of the contour that encircles the upper half ω plane encircles the origin in the D plane. This is the *Nyquist theorem* (cf. Fig. 4).

References for Chapter 1

Birdsall, C. K., G. R. Brewer, and A. V. Haeff: Resistive wall amplifier. Proc. IRE **41**, 865 (1953)

Briggs, R. J.: Electron stream interaction with plasmas. Cambridge, Mass.: MIT 1964

Chu, L. J.: Proc. IRE Conference on electron device. Durham, New Hampshire (1951) (unpublished)

Hasegawa, A.: Theory of longitudinal plasma instabilities. Phys. Rev. **169**, 204 (1968)

Landau, L. D., Lifshitz, E. M.: Electrodynamics of continuous media, p. 253. Reading, Mass.: Pergamon 1960

Sturrock, P. A.: Kinematics of growing waves. Phys. Rev. **112**, 1488 (1958)

Weibel, E. S.: Spontaneously growing transverse waves in a plasma due to an anisotropic velocity distribution. Phys. Rev. Letters **2**, 83 (1959)

2. Microinstabilities – Instabilities Due to Velocity Space Nonequilibrium

2.1 Introduction

In this chapter, we treat instabilities that originate from a non-Maxwellian type velocity distribution. Except for special cases (Section 2.5), we assume a uniform plasma. Because plasmas in space are collision free, they either maintain their original distribution functions or the distribution functions change slowly due to their movement through the background magnetic field. Hence the velocity distribution function is almost never Maxwellian but rather is determined by the past history of the plasma. The velocity distribution can differ from a Maxwellian in basically two different ways. One occurs when the distribution function has more than two humps: for example, when a group of streaming particles is intermixed with the background plasma or when electrons and protons have different average velocities. Such cases are treated in Section 2.2. The other way is by having an anisotropic distribution. The force due to the magnetic field acts only in the direction perpendicular to the field, causing particles with collision frequency much smaller than the cyclotron frequency to move in an anisotropic way. This creates an anisotropic velocity distribution with respect to the direction of the ambient magnetic field. We discuss instabilities associated with such a distribution in Section 2.3. Section 2.4 is devoted to instabilities associated with anisotropic velocity distributions which occur in the hydromagnetic frequency range (frequencies much smaller than the proton cyclotron frequency). In the last section of this chapter, Section 2.5, we treat instabilities in partially ionized plasmas as they relate to various ionospheric phenomena. These instabilities usually originate from the electron-ion two stream effect or from a combination of the two stream and a density gradient. We will see that the existence of a density gradient modifies the two stream instability in a very interesting way in a collision dominated plasma.

2.2 Instabilities Due to Two Humped Velocity Distributions

This is probably the most popular plasma instability. The instability is generated by the fact that the velocity distribution function has two humps. The two stream instability is one class of this instability, because the "stream" creates an additional hump at $v=v_0$ in the distribution function of the background plasma. A current in a plasma also produces two humps in the combined velocity distribution of electrons and ions because the average velocity of electrons is different from that of the ions.

This instability has been applied to the solar corona, the solar wind, the magnetosphere and its boundary and to the ionosphere. One of the most important consequences of this instability is the generation of a large electric field in the direction parallel to the magnetic field which is produced by the electron-ion two stream instability in association with the field aligned current. We will detail this effect in Subsection 2.2d, after presenting the theoretical aspects of the instabilities in Subsections 2.2a, 2.2b and 2.2c. We treat the electrostatic instability in 2.2a, the electromagnetic instability in 2.2b and discuss the effect of the magnetic mirror in 2.2c.

2.2a Electrostatic Instabilities

We have already seen in Section 1.3 that an electron stream existing in a plasma produces a plasma instability. The instability is interpreted as a consequence of the coupling between the negative energy wave in the stream and the plasma wave in the stationary plasma. It was also pointed out that when the stream is collisional, the negative energy nature changes to negative dissipative nature but the instability can still persist.

If the distribution function is not monochromatic, there are particles with velocities close to the phase velocity of the wave which contribute to the dissipation of the wave energy. Hence the instability becomes also the negative dissipative type.

Let us look at such a case. We first consider a two humped *electron distribution* function in the direction of the magnetic field. We must use the Vlasov equation to analyze such a case.

We first consider the case in which the direction of propagation is parallel to the magnetic field. Because we assume an electrostatic perturbation ($\nabla \times \boldsymbol{E}=0$), we can use the electrostatic potential φ to represent the field. The linearized Vlasov equation for the electron distribution function then becomes

$$\frac{\partial f_1}{\partial t} + v\frac{\partial f_1}{\partial z} + \frac{e}{m_e}\frac{\partial \varphi_1}{\partial z}\frac{\partial f_0}{\partial v} = 0, \tag{2.1}$$

where z is the direction of propagation which we have assumed to be also the direction of the background magnetic field. In Eq. (2.1),

f_1 $[=f_1(z, v, t)]$ is the perturbed velocity distribution function of an electron, and f_0 $[=f_0(v)]$ is the unperturbed velocity distribution function that depends only on velocity v. Taking the Fourier-Laplace transformation as defined in Eq. (1.37), we have

$$f_1 = -\frac{(e/m_e)(\partial f_0/\partial v)}{v-\omega/k}\varphi_1 . \tag{2.2}$$

The perturbed charge density qn_1 can then be obtained from Eq. (1.13) as

$$qn_1 = -en_0 \int_{-\infty}^{\infty} f_1\, dv = \frac{e^2 n_0}{m_e} \int_{-\infty}^{\infty} \frac{(\partial f_0/\partial v)\,\varphi_1}{v-\omega/k}\, dv . \tag{2.3}$$

Because we consider here an instability generated by the electron two humped distribution, we ignore the ion contribution to the charge density perturbation. As was discussed in Chapter 1, the Fourier-Laplace transformation is valid for the plane $\operatorname{Im}\omega > 0$, hence the integral over velocity space in the above equation has to go below the pole at ω/k if $k > 0$ (see Fig. 5). The conductivity associated with such an electron

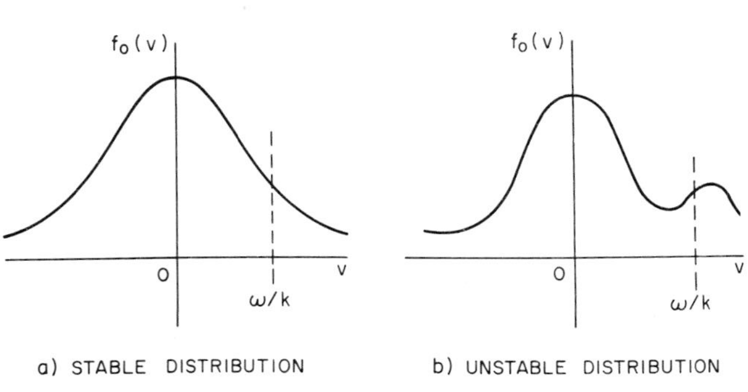

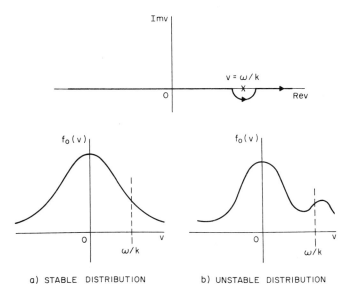

Fig. 5a and b. Path of integration of Eq. (2.3) over velocity v, and the corresponding (a) stable and (b) unstable velocity distributions

perturbation can then be obtained by using Eq. (1.60)

$$\sigma = \frac{i\omega\varepsilon_0\omega_{pe}^2}{k^2} \int_{-\infty}^{\infty} \frac{\partial f_0/\partial v}{v-\omega/k}\, dv\,. \tag{2.4}$$

The dispersion relation is obtained by substituting the above expression into Eq. (1.62)

$$1 - \frac{\omega_{pe}^2}{k^2} \int_{-\infty}^{\infty} \frac{\partial f_0/\partial v}{v-\omega/k}\, dv = 0\,. \tag{2.5}$$

According to the criterion presented in Eq. (1.53), the necessary condition of instability is obtained from the condition Re $\sigma < 0$. If the growth rate is very small, i.e., if Im ω is small, σ in Eq. (2.4) can be expressed in terms of the principal integral and the contribution from the pole at $v = \omega/k$ as (cf. Fig. 5)

$$\sigma = \frac{i\omega\varepsilon_0\omega_{pe}^2}{k^2} \left[P \int_{-\infty}^{\infty} \frac{\partial f_0/\partial v}{v-\omega/k}\, dv + i\pi \frac{k}{|k|} \left.\frac{\partial f_0}{\partial v}\right|_{v=\omega/k} \right] \tag{2.6}$$

where P represents the principal integral. The necessary condition for an instability with a small growth rate is then expressed immediately by

$$\frac{\omega}{k} \left.\frac{\partial f_0}{\partial v}\right|_{v=\omega/k} > 0\,. \tag{2.7}$$

Physically, Eq. (2.7) means that at $v = \omega/k\,(= v_p > 0)$, the unperturbed velocity distribution function has a positive gradient. Since the Maxwell distribution of the form e^{-v^2} always has a negative gradient, it is always stable (Im $\omega < 0$, this is called the Landau damping). While Eq. (2.7) corresponds to the situation where the number of the particles having a larger velocity than the wave phase velocity is more than those having a smaller velocity, which indicates a net loss of the particle kinetic energy to the wave in their resonant interaction. If the distribution has an additional peak at $v > \omega/k$, the necessary condition of the instability is satisfied. Given a two-humped distribution, the existence of a phase velocity in the range of positive gradient can be found only by solving for real ω and k the dispersion relation given by

$$1 - \frac{\omega_p^2}{k^2} P \int_{-\infty}^{\infty} \frac{\partial f_0/\partial v}{v-\omega/k}\, dv = 0\,. \tag{2.8}$$

For further details see Landau (1946) and Jackson (1960). If it happens that Eq. (2.8) gives a complex ω solution, the above argument breaks down. For example, if f_0 represents two streams with no velocity spread, i.e., $f_0=\delta(v)+\delta(v-v_0)$, Eq. (2.8) gives the same dispersion relation shown in the example in Chapter 1, and has a complex ω solution. Then the instability occurs as a result of interaction between negative and positive energy waves, and not by wave-particle interaction, as in Eq. (2.7).

If the growth rate is small (this corresponds to the case where Eq. (2.8) has a real ω solution), we can use expression (1.65) to obtain the growth rate of the instability. From Eq. (2.6), we can see

$$\operatorname{Im} \varepsilon = -\pi \frac{\omega_p^2}{k^2} f_0'\left(\frac{\omega}{k}\right) \frac{k}{|k|} \tag{2.9}$$

and

$$\frac{\partial \operatorname{Re} \varepsilon}{\partial \omega} = -\frac{\omega_p^2}{k^3} P \int_{-\infty}^{\infty} \frac{\partial f_0/\partial v}{(v-\omega/k)^2} dv \tag{2.10}$$

where $f_0'(\omega/k)=\partial f_0/\partial v|_{v=\omega/k}$.

If we use the dispersion relation (2.8), we can express the integral in Eq. (2.10) in terms of the phase (ω/k) and the group $(\partial\omega/\partial k)$ velocity of the wave. Multiplying Eq. (2.8) by k^2 and taking the derivative with respect to k, we obtain

$$\frac{\omega_p^2}{k^2} \int_{-\infty}^{\infty} \frac{\partial f_0/\partial v}{(v-\omega/k)^2} dv = \frac{2}{\partial\omega/\partial k-\omega/k}.$$

Using this expression in Eq. (2.10), we can obtain the growth rate γ as

$$\gamma = -\frac{\operatorname{Im} \varepsilon}{\partial \operatorname{Re} \varepsilon/\partial\omega} = \frac{\pi}{2} \frac{\omega_p^2}{|k|} f_0'\left(\frac{\omega}{k}\right)\left(\frac{\omega}{k}-\frac{\partial\omega}{\partial k}\right). \tag{2.11}$$

Although the above expression is not complete in that the actual growth rate can be obtained only by solving the dispersion relation (2.8) to express ω in term of k, Eq. (2.11) has an interesting property. According to this expression, for an instability to occur $(\gamma>0)$, we need not only to have $f_0'(\omega/k)>0$, but we also need the additional condition that the phase velocity of the wave be larger than the group velocity (Dawson, 1961).

To appreciate this result, we consider waves carried by drifting electrons. The distribution $f_0(v)$ may be symmetric around the drift

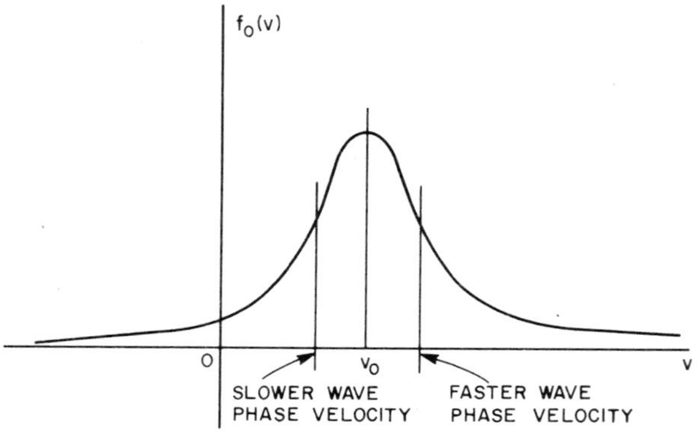

Fig. 6. Wave particle resonance of slower and faster waves on a drifting plasma

speed v_0 as shown in Fig. 6. As discussed in Chapter 1, such a drifting electron will carry two kinds of waves, one having a phase velocity larger than the drift speed v_0 (the faster wave) and the other having a phase velocity smaller than v_0 (the slower wave). The group velocity for both waves is approximately v_0 if the distribution function has a small spread ($v_T < v_0$). Hence the slower wave, for which $f_0'(\omega/k) > 0$, is still a damped wave because $\omega/k < \partial\omega/\partial k$ in Eq. (2.11). This is because the slower wave having a negative energy reduces its amplitude by absorbing energy from the particles. Hence we can see that even though the existence of a positive slope in the distribution function is a necessary condition for an instability, if the positive slope of the velocity distribution function corresponds to the region of the negative energy wave phase velocity, the plasma is stable. For an instability, we *need* a two humped distribution, because only in such a case can a positive energy wave exist with the phase velocity in the positive slope regions!

We now consider a low frequency regime where ion dynamics become important. The dispersion relation including ion dynamics can be obtained in exactly the same manner as above by adding the ion charge density perturbation to the conductivity

$$1 - \frac{\omega_{pe}^2}{k^2} \int_{-\infty}^{\infty} \frac{\partial f_0^{(e)}/\partial v}{v - \omega/k}\, dv - \frac{\omega_{pi}^2}{k^2} \int_{-\infty}^{\infty} \frac{\partial f_0^{(i)}/\partial v}{v - \omega/k}\, dv = 0, \qquad (2.12)$$

where $\omega_{pi}\ (=(e^2 n_0/\varepsilon_0 m_i)^{1/2})$ is the ion plasma (angular) frequency. (The ion and electron distributions are designated by the superscripts i and e.) Although this dispersion relation is not analytically solvable for the general case, we can solve it for two important cases. One is the case where the ions are cold and stationary whereas the electrons are hot with their thermal velocity much larger than the phase velocity and drift with a velocity v_0 with respect to the ions. The other is the case where both electrons and ions are cold. In either case, the integral for the ion contribution in Eq. (2.12) becomes, after integration by parts

$$I_i \equiv \int_{-\infty}^{\infty} \frac{\partial f_0^{(i)}/\partial v}{v-\omega/k}\, dv = \int_{-\infty}^{\infty} \frac{f_0^{(i)}}{(v-\omega/k)^2}\, dv .$$

The "cold" distribution can be represented by the delta function as

$$f_0^{(i)} = \delta(v) . \tag{2.13}$$

We first consider the former case, the case with hot electrons, where

$$\begin{aligned} I_e &\equiv \int_{-\infty}^{\infty} \frac{\partial f_0^{(e)}/\partial v}{v-\omega/k}\, dv \\ &= P\int_{-\infty}^{\infty} \frac{\partial f_0^{(e)}/\partial v}{v-\omega/k}\, dv + i\pi \frac{k}{|k|} \left.\frac{\partial f_0^{(e)}}{\partial v}\right|_{v=\omega/k} \end{aligned} \tag{2.14a}$$

Because the thermal velocity for electrons is assumed much larger than the phase velocity, we can assume a drifting Maxwellian distribution for $f_0^{(e)}$ such that

$$f_0^{(e)} = \frac{1}{(2\pi)^{1/2} v_{Te}}\, e^{-(v-v_0)^2/2v_{Te}^2} .$$

Then,

$$\begin{aligned} I_e &\cong \frac{1}{(2\pi)^{1/2} v_{Te}} \int_{-\infty}^{\infty} \frac{1}{v} \cdot \left(-\frac{v}{v_{Te}^2}\, e^{-v^2/2v_{Te}^2}\right) dv \\ &\quad + i\pi \frac{k}{|k|} \left.\frac{\partial f_0^{(e)}}{\partial v}\right|_{v=\omega/k} \\ &= -\frac{1}{v_{Te}^2} + i\pi \frac{k}{|k|} \left.\frac{\partial f_0^{(e)}}{\partial v}\right|_{v=\omega/k} \end{aligned} \tag{2.14b}$$

where we assume the electron drift velocity v_0 to be much smaller than the thermal velocity v_{Te}. The dispersion relation Eq. (2.12) becomes

$$1 - \frac{\omega_{pi}^2}{\omega^2} + \frac{\omega_{pe}^2}{k^2 v_{Te}^2} - i\pi \frac{\omega_{pe}^2}{k^2} \frac{k}{|k|} \left.\frac{\partial f_0^{(e)}}{\partial v}\right|_{v=\omega/k} = 0 . \tag{2.15}$$

For a low frequency limit, $\omega \ll \omega_{pi}$, the real part of the dispersion relation is readily solvable

$$\omega = k c_s, \tag{2.16}$$

where c_s is the *ion sound speed* given by

$$c_s = v_{Te}(m_e/m_i)^{1/2}. \tag{2.17}$$

The wave represented by the dispersion relation (2.17) is called the *ion acoustic (sound) wave*. The growth rate γ of the instability is obtained using Eq. (1.65)

$$\gamma = \frac{\pi}{2}\,\omega v_{Te}^2 \left.\frac{\partial f_0^{(e)}}{\partial v}\right|_{v=c_s}. \tag{2.18a}$$

As expected, the instability ($\gamma > 0$) occurs when $\partial f_0^{(e)}/\partial v$ is positive at $v = c_s$. This means the drift velocity v_0 of the electrons has to be greater than c_s because $f_0^{(e)}$ has its only peak at $v = v_0$.

When $v_0 \gtrsim \omega/k$, the growth rate γ can be calculated using Eq. (2.18a)

$$\gamma \sim \sqrt{\frac{\pi}{8}}\,\omega\,\frac{v_0}{v_{Te}} \sim \sqrt{\frac{\pi}{8}}\,\omega\,\sqrt{\frac{m_e}{m_i}}. \tag{2.18b}$$

We have shown that when the ions are cold and the electrons are hot the ion sound wave becomes unstable when the electron drift velocity exceeds the ion sound speed. However, if the ion temperature is comparable with the electron temperature, the instability becomes possible only when the electron drift speed exceeds the electron thermal speed, v_{Te}, which is 43 $[=(m_i/m_e)^{1/2}]$ times larger than c_s (Fried and Gould, 1961).

The growth rate obtained in Eq. (2.18) is very small because $\gamma/\omega \sim \sim (m_e/m_i)^{1/2}$. We conclude that when $v_0 < v_{Te}$, the instability is generated by a small number of resonant electrons with velocities comparable to the phase velocity of the wave. In the other case, where both electrons and ions are cold, $v_0 > v_{Te}$, and we will see that the instability becomes more violent. This is because instead of a small number of resonant electrons, the entire electron distribution contributes to the instability. Such a case was treated by Pierce (1948) and later by Buneman (1958) and is often called the Buneman instability. The nature of the instability is quite similar to the electron two stream instability introduced in Chapter 1. The dispersion relation is obtained from Eq. (2.12) by assuming an electron velocity distribution of $\delta(v - v_0)$:

$$1 - \frac{\omega_{pi}^2}{\omega^2} - \frac{\omega_{pe}^2}{(\omega - k v_0)^2} = 0. \tag{2.19}$$

An unstable root of this dispersion relation can be obtained by using the fact that the instability is generated by the slower wave on the electron stream: $\omega = k v_0 - \omega_{pe}$. If we multiply Eq. (2.19) by the common denominator and divide by $\omega - k v_0 - \omega_{pe}$ (because the fast wave, $\omega = k v_0 + \omega_{pe}$, is uncoupled) we have

$$(\omega - k v_0 + \omega_{pe})\omega^2 = \frac{\omega_{pi}^2 (\omega - k v_0)^2}{\omega - k v_0 - \omega_{pe}}.$$

We then choose a wave number $k \sim \omega_{pe}/v_0$ and assume $\omega \ll \omega_{pe}$, to obtain

$$\omega^3 \sim -\frac{\omega_{pe}^3}{2} \frac{m_e}{m_i}.$$

One of the roots of this equation has a positive imaginary part showing an unstable solution. The growth rate γ is thus obtained as

$$\gamma \sim \left(\frac{m_e}{m_i}\right)^{1/3} \omega_{pe}. \tag{2.20}$$

If we compare this growth rate with that of Eq. (2.18b), we can see that the growth rate for the case of a cold electron stream (when $v_0 > v_{Te}$) is much larger than in the case with a hot electron stream with $v_0 < v_{Te}$. As was indicated before, this is because the instability due to a cold electron stream is generated by the contribution of the entire group of streaming electrons.

We have seen that a current in a plasma generates instabilities for a wave propagating *parallel* to the direction of the current. If we are to apply these results to real phenomena, such as to the field aligned current in the topside ionosphere, we must ask ourselves whether these are the only possible instabilities. Certainly we have not explored all the possible cases. What about a wave propagating obliquely to the current, or what about an electromagnetic perturbation? We must check whether these cases produce either a smaller threshold, or a larger growth rate than the cases treated above.

Leaving the electromagnetic case to the next subsection, let us consider the effect of an obliquely propagating wave. At first sight, such a wave seems more stable because the effect of the electron drift is reduced by the cosine factor of the angle between the $\boldsymbol{k}$ vector of the wave and the direction of the drift. However, the existence of a background magnetic field allows the propagation of a set of new waves called electrostatic ion cyclotron waves (obliquely propagating ion Bernstein waves) whose damping increment is small even if the ion temperature T_i is comparable to the electron temperature T_e. This is in contrast to the ion acoustic wave in which a heavy damping arises when T_i becomes comparable to T_e. Therefore we may expect, particularly when $T_i \sim T_e$, that the ion cyclotron wave will become unstable at a smaller electron drift speed. This instability was first discussed by Drummond and Rosenbluth (1962) and was carefully studied by Kindel and Kennel (1971) for its application to the topside current instability. Hence, let us now look at the instability of ion cyclotron waves.

Because the current flows in the direction parallel to the magnetic field, the oblique wave we are going to study has its wave vector component perpendicular to the ambient magnetic field. In such a case, the solution of the Vlasov equation becomes much more complicated because of the effect of cyclotron motion. The position of a particle perpendicular to the magnetic field may be given by $x_\perp \sim (v_\perp \sin \omega_c t)/\omega_c$, where $v_\perp$ is the perpendicular velocity of the particle. Hence the electromagnetic field, say $E \exp i(\boldsymbol{k} \cdot \boldsymbol{x} - \omega t)$, seen by the particle whose location is changing periodically may be modified to $E \exp i\left[\frac{k_\perp v_\perp}{\omega_c} \sin(\omega_c t) + k_{||} v_{||} t - \omega t\right] \sim E \sum_{n=-\infty}^{\infty} J_n\left(\frac{k_\perp v_\perp}{\omega_c}\right) \exp i(n\omega_c + k_{||} v_{||} - \omega) t$, where $k_\perp$ and $k_{||}$ are wave numbers projected perpendicular and parallel to the direction of the magnetic field. If we average over the velocities of all the particles, this effect produces a charge density perturbation which has the cyclotron harmonic structure. The wave generated by this effect was first found by Bernstein (1958) and is often called a *Bernstein wave*. The ion cyclotron wave instability we are going to discuss now is an instability of the obliquely propagating ion Bernstein wave generated by electrons drifting parallel to the magnetic field.

In the presence of a magnetic field, the Vlasov equation for the unperturbed velocity distribution function f_0 becomes

$$(\boldsymbol{v} \times \boldsymbol{B}_0) \cdot \frac{\partial f_0}{\partial \boldsymbol{v}} = 0. \tag{2.21}$$

Eq. (2.21) can be satisfied for any function of $v_\perp$ and $v_{||}$ as

$$f_0(\boldsymbol{v}) = f_0(v_\perp, v_{||}) \tag{2.22}$$

where $v_{||}(=v_z)$ and $v_\perp = (v_x^2 + v_y^2)^{1/2}$ are the velocities parallel and perpendicular to the magnetic field, and z is taken to be parallel to $\boldsymbol{B}_0$. As has been pointed out, the thermodynamic equilibrium situation is achieved when f_0 is the Maxwell distribution; hence, any anisotropic distribution or two-humped distribution either in $v_\perp$ or $v_{||}$ produces free energy in the plasma and may cause instabilities.

If we again limit our interest to electrostatic perturbations,* the linearized Vlasov equation for a species with charge q and mass m becomes

$$\frac{\partial f_1}{\partial t}+\boldsymbol{v}\cdot\frac{\partial f_1}{\partial \boldsymbol{x}}+\frac{q}{m}(\boldsymbol{v}\times\boldsymbol{B}_0)\cdot\frac{\partial f_1}{\partial \boldsymbol{v}}=\frac{q}{m}\frac{\partial \varphi_1}{\partial \boldsymbol{x}}\cdot\frac{\partial f_0}{\partial \boldsymbol{v}}. \qquad (2.23)$$

In Eq. (2.23) f_1 can be obtained either by integrating along the unperturbed orbit (for example, see Stix, 1962) or by solving the differential equation by using cylindrical coordinates in velocity space (for example, see Bernstein, 1958). We take the former method. If we use the trajectory decided by the stationary field i.e., $\ddot{\boldsymbol{x}}=q/m\,(\boldsymbol{v}\times\boldsymbol{B}_0)$, the left hand side of Eq. (2.23) can be written as a total derivative with respect to time, df_1/dt. f_1 can then be integrated formally along such a trajectory

$$f_1(\boldsymbol{x},\upsilon,t)=\int_{-\infty}^{t} dt'\,\frac{q}{m}\frac{\partial \varphi_1(t')}{\partial \boldsymbol{x}'}\cdot\frac{\partial f_0}{\partial \upsilon}\bigg|_{\boldsymbol{v}=\boldsymbol{v}'(t')} \qquad (2.24)$$

where, the unperturbed trajectory is given by

$$\begin{aligned}
x'(t')-x&=\frac{\upsilon_\perp}{\omega_c}\{\sin[\omega_c(t'-t)+\theta]-\sin\theta\},\\
y'(t')-y&=\frac{\upsilon_\perp}{\omega_c}\{\cos[\omega_c(t'-t)+\theta]-\cos\theta\}, \qquad (2.25)\\
z'(t')-z&=\upsilon_{\|}(t'-t).
\end{aligned}$$

The final position at $t'=t$ is chosen to be $x'=x$, $y'=y$ and $z'=z$, while the corresponding velocities are $\upsilon_x(t)=\upsilon_\perp\cos\theta$, $\upsilon_y(t)=-\upsilon_\perp\sin\theta$ and $\upsilon_z(t)=\upsilon_{\|}$. Note that the sense of rotation is left (right) handed with respect to the direction of the magnetic field for protons (electrons) where $\omega_c=\omega_{ci}=eB_0/m_i>0\,(\omega_c=-\omega_{ce}=-eB_0/m_e<0)$ is the cyclotron frequency.

If we consider a perturbation of the form $\exp i(\boldsymbol{k}\cdot\boldsymbol{x}-\omega t)$ and take the direction of wave propagation in the x, z plane (without loss of generality) such that

$$\boldsymbol{k}\cdot\boldsymbol{x}=k_\perp x+k_{\|}z, \qquad (2.26)$$

* Although the previous analysis with $k_\perp=0$ is exactly electrostatic, in the present case this is an approximation. The electromagnetic effect becomes important for a large $\beta\,(\gtrsim 10^{-2})$ plasma, where β is the ratio of the plasma pressure to the magnetic pressure.

Eq. (2.24) gives

$$f_1(\boldsymbol{x}, \boldsymbol{v}, t) = \frac{iq}{m} \int_{-\infty}^{t} dt' \left(k_\perp \frac{\partial f_0}{\partial v_x} + k_{||} \frac{\partial f_0}{\partial v_{||}} \right) \Bigg|_{\boldsymbol{v}=\boldsymbol{v}'(t')} \cdot \varphi_1 \exp i[k_\perp x'(t') + k_{||} z'(t') - \omega t']. \tag{2.27}$$

If we substitute x' and z' of Eq. (2.25) into the above expression and note that $\partial f_0/\partial v_{||}\,|_{v_{||}=v'_{||}} = \partial f_0/\partial v_{||}$ (because $v_{||}$ does not change), while $\partial f_0/\partial v_x\,|_{v_x=v'_x} = (\partial f_0/\partial v_\perp)\,(v_\perp/v_x)\,|_{v_x=v'_x} = \partial f_0/\partial v_\perp \cos[\omega_c(t'-t)+\theta]$ (because $v_\perp$ is also a constant of motion), we can integrate Eq. (2.27). We then obtain the perturbed distribution function f_1 as a function of *independent* variables, $\boldsymbol{x}$, $\boldsymbol{v}$ and t. If we then integrate over the velocity space to obtain the charge density perturbation, the equivalent conductivity is obtained as

$$\sigma(\omega, k) = \frac{i\omega\varepsilon_0\omega_p^2}{k^2} \sum_{\mathrm{n}=-\infty}^{\infty} \int d\boldsymbol{v} \left[J_{\mathrm{n}}\left(\frac{k_\perp v_\perp}{\omega_c}\right)\right]^2 \frac{k_{||}(\partial f_0/\partial v_{||}) + (\mathrm{n}\,\omega_c/v_\perp)(\partial f_0/\partial v_\perp)}{k_{||}v_{||} - (\omega - \mathrm{n}\,\omega_c)} \tag{2.28}$$

where $d\boldsymbol{v} = 2\pi v_\perp\, dv_\perp\, dv_{||}$ and J_{n} is the nth order Bessel function of the first kind. In deriving Eq. (2.28) use is made of the identity

$$\exp(iz\sin\theta) = \sum_{\mathrm{n}=-\infty}^{\infty} J_{\mathrm{n}}(z)\exp(i\mathrm{n}\theta).$$

To study the instability, it is convenient to specify the actual form of the distribution function $f_0(v_\perp, v_{||})$. If we take a frame where ions are stationary, the distribution function of ions, $f_0^{(i)}$, may be assumed to be an isotropic Maxwellian*;

$$f_0^{(i)}(v) = \left(\frac{1}{\sqrt{2\pi}\, v_{Ti}}\right)^3 e^{-\frac{v_\perp^2 + v_{||}^2}{2v_{Ti}^2}} \tag{2.29}$$

The ion conductivity $\sigma_i(\omega, k)$ is then obtained from Eq. (2.28) as

$$\sigma_i(\omega, k) = \frac{-i\omega\varepsilon_0\omega_{pi}^2}{k^2 v_{Ti}^2} \sum_{\mathrm{n}=-\infty}^{\infty} e^{-\lambda_i} I_{\mathrm{n}}(\lambda_i) \left[1 + \frac{\omega}{\sqrt{2}\,k_{||}v_{Ti}} Z\left(\frac{\omega - \mathrm{n}\,\omega_{ci}}{\sqrt{2}\,k_{||}v_{Ti}}\right)\right], \tag{2.30}$$

---

* We treat the effect of anisotropic distribution in Section 2.3.

where

$$\lambda_i = \frac{k_\perp^2 v_{Ti}^2}{\omega_{ci}^2}.$$

I_n is the modified Bessel function of the nth order and Z is the plasma dispersion function (Fried and Conte, 1961) defined by

$$Z(\zeta) = \frac{1}{(\pi)^{1/2}} \int_{-\infty}^{\infty} \frac{e^{-x^2}}{x-\zeta} dx \quad \text{for} \quad \text{Im}\, \zeta > 0 \tag{2.31}$$

$$= \text{analytic continuation of the above integral for Im}\, \zeta < 0.$$

The power-series expansion for a small argument and the asymptotic expansion for a large argument of the Z function are given by

$$Z(\zeta) \cong i(\pi)^{1/2} e^{-\zeta^2} - 2\zeta\left(1 - \frac{2\zeta^2}{3}\right) \quad \text{for } |\zeta| \ll 1 \tag{2.32}$$

$$Z(\zeta) \cong i(\pi)^{1/2} \sigma e^{-\zeta^2} - \frac{1}{\zeta}\left(1 + \frac{1}{2\zeta^2}\right) \quad \text{for } |\zeta| \gg 1 \tag{2.33}$$

where

$$\sigma = \begin{cases} 0 & \text{Im}\, \zeta > 0 \\ 1 & \text{Im}\, \zeta = 0 \\ 2 & \text{Im}\, \zeta < 0. \end{cases}$$

For electrons, if we assume a drift Maxwellian as given by the first line in Eq. (2.14b), the conductivity may be simplified for $\omega \ll \omega_{ce}$, to

$$\sigma_e(\omega, k) = \frac{-i\omega\varepsilon_0\omega_{pe}^2}{k^2 v_{Te}^2}\left[1 + \frac{\omega - k_\parallel v_0}{\sqrt{2}\, k_\parallel v_{Te}} Z\left(\frac{\omega - k_\parallel v_0}{\sqrt{2}\, k_\parallel v_{Te}}\right)\right]. \tag{2.34}$$

Thus the dispersion relation is finally obtained as

$$\sigma_i \text{ (Eq. 2.30)} + \sigma_e \text{ (Eq. 2.34)} - i\omega\varepsilon_0 = 0. \tag{2.35}$$

If we look at the ion conductivity, Im Z is always positive, therefore, Re σ_i is always positive which indicates a dissipative effect. The damping which corresponds to the dissipation from the $n=0$ term is the ion Landau damping and that from the $n \geq 1$ terms is called cyclotron damping.

On the other hand the real part of the electron conductivity $\mathrm{Re}\,\sigma_e$ can become negative if $v_0 > \omega/k$, the phase velocity of the wave. Hence the drifting electrons contribute to the negative dissipation and the instability occurs when the electron negative dissipation overcomes the ion Landau or cyclotron damping.

Let us see now why the ion cyclotron wave instability for $T_e \sim T_i$ requires a smaller threshold on the drift velocity of electrons than the ion acoustic wave instability. For an ion acoustic wave with a finite ion temperature, we cannot use Eq. (2.16), but must evaluate Eq. (2.12) directly. If the ions have a Maxwellian distribution, the corresponding ion conductivity may be obtained from Eq. (2.30) by putting $\lambda_i = 0$ (because $k_\perp = 0$ for the ion acoustic wave to be present) as

$$\sigma_i(\omega, k) = -\frac{i\omega\varepsilon_0\omega_{pi}^2}{k^2 v_{Ti}^2}\left[1 + \frac{\omega}{\sqrt{2}\,k_{\|} v_{Ti}} Z\left(\frac{\omega}{\sqrt{2}\,k_{\|} v_{Ti}}\right)\right]. \tag{2.36}$$

When $T_e \sim T_i$, the phase velocity of the ion acoustic wave, $v_{Te}(m_e/m_i)^{1/2}$, approaches the ion thermal speed v_{Ti}. Then the imaginary part of the Z function in the above expression reaches its maximum and hence heavy ion Landau damping results. This makes the ion acoustic wave difficult to excite and results in an increase in the electron threshold drift speed to v_{Te}.

On the other hand, in the case of the ion cyclotron wave instability, by taking a small $k_\| v_{Ti}$, such that $\omega - \omega_{ci} \gg k_\| v_{Ti}$, and a large perpendicular wave number $k_\perp$, we can have a wave with a relatively small phase velocity $(\omega/k_\|)$ due to the ion cyclotron resonance. This reduces the threshold drift speed of electrons necessary to excite the instability. Fig. 7 shows the comparison of the threshold of the electron drift speed necessary to excite the ion acoustic wave and the ion cyclotron wave obtained by Kindel and Kennel (1971). We can see that there is a wide range of T_e/T_i where the threshold drift speed of the ion cyclotron instability is less than that of the ion acoustic instability.

We have seen how a current parallel to the magnetic field can cause a variety of instabilities. When $v_0 < v_{Te}$, an ion acoustic wave is excited if $k_\perp = 0$ and an ion cyclotron wave is excited when $k_\perp \neq 0$, while if $v_0 > v_{Te}$, the electron-ion two stream (Buneman) instability is excited. These instabilities tend to block the current by dissipating the current energy into the electric field energy, thus producing a finite parallel electric field. When the current is very weak or when $T_e \sim T_i$, the ion cyclotron instability may be excited. When the current is increased, the growth rate becomes larger than the ion cyclotron frequency. Then the periodic gyromotion of the ions is destroyed and the ions become demagnetized. When this happens, the ion cyclotron wave is lost and the instability

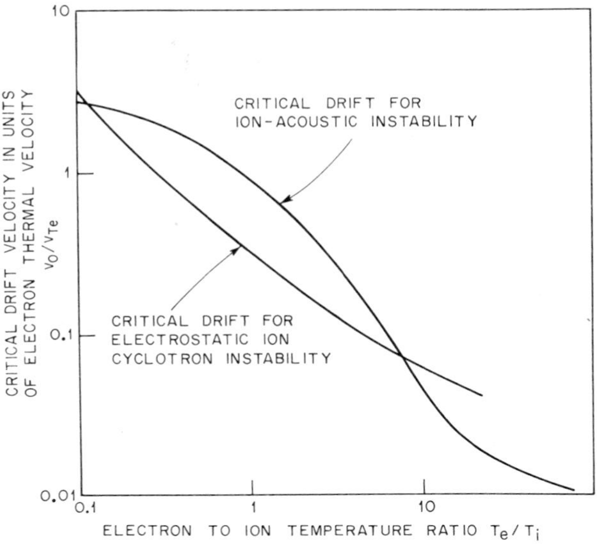

Fig. 7. Hydrogen plasma critical drifts, normalized to the electron thermal speed, of the ion cyclotron and ion acoustic waves as a function of the electron to ion temperature ratio, T_e/T_i. The H^+ cyclotron wave is unstable to the smaller current over the range $0.1 < T_e/T_i < 8$. (After Kindel and Kennel, 1970)

changes its nature to the ion acoustic type even if $k_\perp \neq 0$. If the current is further increased, the instability changes to the Buneman type.

We must look at the data carefully to see which range of the phenomena to expect. Perhaps the most interesting application of these instabilities is the field aligned current observed during magnetospheric substorms. We will discuss this problem in Subsection 2.2d. Here we will look at applications to other phenomena. One obvious region in which "two streams" of plasma can exist is where the solar wind intersects with planetary plasmas. One of the earliest papers to suggest such a two stream instability is the one by Gintsburg (1960), who concludes that a solar corpuscular stream of 3×10^4 °K with $v_0 = 10^6$ m/s and a number density of 10/cm^3 excites a plasma oscillation in the exosphere ($h \sim 2000$ km). This estimate seems rather unrealistic because of the existence of the magnetosphere (which would prohibit the penetration of such a stream), however, it may be applicable to other planets and suggests an instability at the earth's magnetopause.

In 1965, Scarf *et al.* (1965) suggested that the ion acoustic wave instability might exist at the transition region of the solar wind and the magnetosphere.

The solar wind has an average speed of about 400 km/sec. Therefore if the ion acoustic speed in the plasma in the transition region is less than this speed, and if $T_e \gg T_i$ one can expect the instability. This means the electron temperature T_e(eV) at the magnetopause should satisfy

$$c_s \equiv v_{Te}\left(\frac{m_e}{m_i}\right)^{1/2} = \left(\frac{eT_e}{m_i}\right)^{1/2} < 4 \times 10^5 \text{ m/s},$$

or

$$T_e < 1.6 \text{ keV}.$$

Existence of electric field intensities up to 10 mV/m in the kHz frequency range was observed later on the OGO-5 satellite at the transition region as shown in Fig. 8 (Fredericks *et al.* (1970). Because the ion plasma frequency ω_{pi} is given approximately by $\sqrt{n(\text{m}^{-3})}$ (sec^{-1}), if the plasma density at the magnetopause is 10 cm^{-3}, $\omega_{pi} \sim 3 \times 10^3$, or $f_{pi} = \omega_{pi}/2\pi \sim$ ~ 0.5 kHz. Hence the observed frequency is somewhat higher than ω_{pi} since the ion acoustic wave frequency is smaller than ω_{pi}, this makes the interpretation difficult.

Because the Buneman instability occurs at $\omega \sim \omega_{pe}(m_e/m_i)^{1/3} \sim 3.4\,\omega_{pi}$, the instability frequency lies in the observed frequency range (<3 kHz). However, to excite the Buneman two stream instability, $v_0 > v_{Te}$ must be satisfied. This means the electron temperature must be less than 1 eV in the transition region.

As pointed out by the authors, one important possibility is the effect of Doppler shifting. If the magnetopause is moving at a fraction of the speed of the solar wind, say v_m, the satellite (which is moving at ~ 2 km/s) will see a frequency Doppler shifted by the amount of kv_m. The maximum wave number k is limited by the Debye wave number $k_D = \omega_{pe}/v_{Te}$. Hence the observed frequency could be Doppler shifted by $\omega_{pe}(v_m/v_{Te})$ which may become comparable to ω_{pi}.

On the other hand, we cannot ignore the possibility of other kinds of instabilities. As will be shown in the rest of this chapter there are many other instabilities that can produce electric field oscillations. Instabilities due to an anisotropic velocity distribution or electromagnetic two stream instabilities cannot be ruled out. Even in the framework of electrostatic two stream instabilities, there still exists one important candidate which we have not discussed. That is the *beam cyclotron instability*, which is generated by a relative drift of electrons and ions perpendicular to the magnetic field. Such a drift is possible either due to a difference in collision frequency between electrons and ions and a third group of particles or due to a nonuniformity of the plasma which deforms the periodic ion cyclotron motions into a quasi-straight orbit. In this case the instability is generated even when the relative drift speed v_0 is much

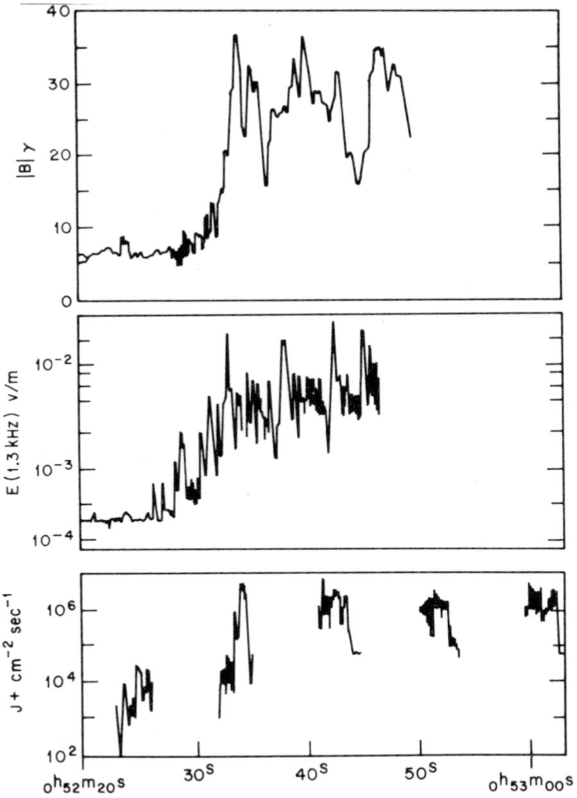

Fig. 8. An example of a sharp well-defined bow shock encountered by OGO-5 during an inbound pass on March 12, 1968. Note the size and the shape of the electric field oscillation at 1.3 kHz and its correlation with the shape of the magnetic field profile $|\boldsymbol{B}|$ and the ion flux variation $\boldsymbol{J}_+$. (After Fredricks *et al.*, 1970)

less than the electron thermal speed (Hasegawa, 1965; Forslund *et al.*, 1970; Krall and Liewer, 1972), and the generated frequency is either the ion acoustic frequency or the electron cyclotron harmonics depending whether the drifting species is electrons or ions. In the latter case the excited frequency ω is $n\omega_{ce}$, where $n=1, 2, 3 \ldots \sim \omega_{pe}/\omega_{ce}$ and $\omega_{ce} \sim \sim 1.8 \times 10^2\ B(\gamma)\ \mathrm{sec}^{-1} \sim 4 \times 10^3\ \mathrm{sec}^{-1}$ for $B_0 \sim 20\ \gamma$. Thus the fundamental cyclotron frequency is approximately 700 Hz and its harmonics can lie very easily in the observed frequency range.

As can be seen from this example, it is often difficult to attribute one instability to one observation; this is often due to a lack of measurable parameters, particularly in spatial correlations.

2.2 b Electromagnetic Instabilities

We will now investigate electromagnetic instabilities excited by a two humped velocity distribution in the direction of the magnetic field. Here "electromagnetic" means a divergence-less electric field, i.e., $\nabla \cdot \boldsymbol{E} \sim 0$, in contrast to the electrostatic case where $\nabla \times \boldsymbol{E} \sim 0$. In reality, however, waves in a hot plasma propagating obliquely to the ambient magnetic field satisfy neither $\nabla \times \boldsymbol{E}=0$ nor $\nabla \cdot \boldsymbol{E}=0$. Only when the $\boldsymbol{k}$ vector is exactly parallel or perpendicular to the magnetic field do the electrostatic waves and electromagnetic waves separate.

In the general case of oblique propagation one has to derive the dispersion relation from the full Maxwell's equation

$$\boldsymbol{k}(\boldsymbol{k} \cdot \boldsymbol{E}_1) - k^2 \boldsymbol{E}_1 + \frac{\omega^2}{c^2} (\boldsymbol{I} + \boldsymbol{\varepsilon}) \cdot \boldsymbol{E}_1 = 0, \tag{2.37}$$

where $\boldsymbol{I}$ is a unit tensor given by

$$\boldsymbol{I} = \begin{pmatrix} 1 & 0 & 0 \\ 0 & 1 & 0 \\ 0 & 0 & 1 \end{pmatrix} \tag{2.38}$$

and $\boldsymbol{\varepsilon}$ is the equivalent dielectric (susceptibility) tensor that is related to the conductivity tensor by

$$\boldsymbol{\varepsilon} = -\frac{\boldsymbol{\sigma}}{i\omega\varepsilon_0} \tag{2.39}$$

with

$$\begin{aligned} \boldsymbol{J}_1 &\equiv q n_0 \int \boldsymbol{v} f_1 \, d\boldsymbol{v} \\ &\equiv \boldsymbol{\sigma} \cdot \boldsymbol{E}_1 . \end{aligned} \tag{2.40}$$

The perturbed distribution function f_1 can be obtained as a function of $\boldsymbol{E}_1$ by solving the Vlasov equation including the full electromagnetic field

$$\frac{\partial f_1}{\partial t} + \boldsymbol{v} \cdot \frac{\partial f_1}{\partial \boldsymbol{x}} + \frac{q}{m} (\boldsymbol{v} \times \boldsymbol{B}_0) \cdot \frac{\partial f_1}{\partial \boldsymbol{v}} = -\frac{q}{m} \left[\boldsymbol{E}_1 + \frac{\boldsymbol{v} \times (\boldsymbol{k} \times \boldsymbol{E}_1)}{\omega} \right] \cdot \frac{\partial f_0}{\partial \boldsymbol{v}} . \tag{2.41}$$

If we use the unperturbed trajectory obtained previously in Eq. (2.25), the left hand side of Eq. (2.41) can again be written as df_1/dt. The perturbed distribution function f_1 is then obtained by integrating the right hand side along the trajectory. The integration is rather tedious, so we skip the detail and show the equivalent dielectric constant obtained by substituting the result into Eqs. (2.39) and (2.40),

$$\varepsilon = -\sum_{\text{species}} \frac{\omega_p^2}{\omega^2}\left\{ \boldsymbol{I} + \sum_{\mathrm{n}=-\infty}^{\infty} \int d\boldsymbol{v}\, \frac{k_{\|}(\partial f_0/\partial v_{\|}) + (\mathrm{n}\,\omega_c/v_\perp)\,(\partial f_0/\partial v_\perp)}{k_{\|} v_{\|} - (\omega - \mathrm{n}\,\omega_c)}\, \boldsymbol{S} \right\} \tag{2.42}$$

where $\boldsymbol{I}$ is the unit matrix given by Eq. (2.38) and the matrix $\boldsymbol{S}$ is given by

$$\begin{array}{c} \\ \boldsymbol{S} = \end{array} \begin{array}{c} \begin{matrix} \hat{x} \qquad\qquad\qquad & \hat{y} \qquad\qquad\qquad & \hat{z} \end{matrix} \\ \begin{bmatrix} \left(\dfrac{\mathrm{n}\,\omega_c}{k_\perp} J_\mathrm{n}\right)^2 & i\,\dfrac{\mathrm{n}\,\omega_c}{k_\perp} v_\perp J_\mathrm{n} J'_\mathrm{n} & \dfrac{\mathrm{n}\,\omega_c}{k_\perp} v_{\|} J_\mathrm{n}^2 \\ -i\,\dfrac{\mathrm{n}\,\omega_c}{k_\perp} v_\perp J_\mathrm{n} J'_\mathrm{n} & (v_\perp J'_\mathrm{n})^2 & -i v_\perp v_{\|} J_\mathrm{n} J'_\mathrm{n} \\ \dfrac{\mathrm{n}\,\omega_c}{k_\perp} v_{\|} J_\mathrm{n}^2 & i v_\perp v_{\|} J_\mathrm{n} J'_\mathrm{n} & (v_{\|} J_\mathrm{n})^2 \end{bmatrix} \end{array} \begin{matrix} \hat{x} \\ \hat{y} \\ \hat{z}\,. \end{matrix} \tag{2.43}$$

The argument of the Bessel functions J_n is $k_\perp v_\perp/\omega_c$; $\omega_c\ (=qB_0/m)$ is the cyclotron frequency with sign included. In Eq. (2.43), $\boldsymbol{B}_0$ is taken in the direction of positive z axis. The dispersion relation obtainable from Eqs. (2.37) and (2.42) above represents the most general case for a uniform plasma in a uniform magnetic field. For example, the ion cyclotron instability presented in the previous subsection can be derived in a more exact form using the above expression. In fact when β is large $(\geq (m_e/m_i)^{1/2})$, a significant modification is shown to appear. We will not, however, go into the detail of the electromagnetic modification of this instability, but will focus our attention here on instabilities that are predominantly electromagnetic. For this purpose, we first look at the case where the wave propagates parallel to the magnetic field.

In Maxwell's equation (2.37) we put $\boldsymbol{k}\cdot\boldsymbol{E}_1=0$. If we also put $k_\perp=0$ in the dielectric tensor in Eq. (2.43), ε_{yy} becomes equal to ε_{xx}, and ε_{yz} as well as ε_{xz} vanishes. We then have for a transverse wave propagating parallel to the magnetic field

$$\begin{aligned} -k^2 E_x + \frac{\omega^2}{c^2}\,[(1+\varepsilon_{xx})\,(E_x + \varepsilon_{xy} E_y)] &= 0\,, \\ -k^2 E_y + \frac{\omega^2}{c^2}\,[(1+\varepsilon_{xx})\,(E_y - \varepsilon_{xy} E_x)] &= 0\,. \end{aligned} \tag{2.44}$$

If we define a new electric-field vector

$$E_R = E_x - iE_y, \tag{2.45}$$

for the right-hand circularly polarized (RHP) wave and

$$E_L = E_x + iE_y, \tag{2.46}$$

for the left-hand circularly polarized (LHP) wave, E_R and E_L satisfy

$$\left[-k^2 + \frac{\omega^2}{c^2}(1+\varepsilon_{xx}) + i\frac{\omega^2}{c^2}\varepsilon_{xy}\right] E_R = 0 \tag{2.47}$$

and

$$\left[-k^2 + \frac{\omega^2}{c^2}(1+\varepsilon_{xx}) - i\frac{\omega^2}{c^2}\varepsilon_{xy}\right] E_L = 0. \tag{2.48}$$

From Eqs. (2.47) and (2.48), together with the dielectric tensor in Eq. (2.42), we have the dispersion relation for the RHP wave

$$\begin{aligned} k^2c^2 - \omega^2 &+ \omega_{pe}^2 \int \frac{(\omega - kv_{||})f_{0||}^{(e)}(v_{||}) - k(\langle v_{\perp e}^2\rangle/2)(\partial f_{0||}^{(e)}/\partial v_{||})}{\omega - kv_{||} - \omega_{ce}}\, dv_{||} \\ &+ \omega_{pi}^2 \int \frac{(\omega - kv_{||})f_{0||}^{(i)}(v_{||}) - k(\langle v_{\perp i}^2\rangle/2)(\partial f_{0||}^{(i)}/\partial v_{||})}{\omega - kv_{||} + \omega_{ci}}\, dv_{||} = 0, \end{aligned} \tag{2.49}$$

and for the LHP wave

$$\begin{aligned} k^2c^2 - \omega^2 &+ \omega_{pe}^2 \int \frac{(\omega - kv_{||})f_{0||}^{(e)}(v_{||}) - k(\langle v_{\perp e}^2\rangle/2)(\partial f_{0||}^{(e)}/\partial v_{||})}{\omega - kv_{||} + \omega_{ce}}\, dv_{||} \\ &+ \omega_{pi}^2 \int \frac{(\omega - kv_{||})f_{0||}^{(i)}(v_{||}) - k(\langle v_{\perp i}^2\rangle/2)(\partial f_{0||}^{(i)}/\partial v_{||})}{\omega - kv_{||} - \omega_{ci}}\, dv_{||} = 0. \end{aligned} \tag{2.50}$$

In the above expressions, $f_{0||}^{(e)}$ and $f_{0||}^{(i)}$ are the electron and ion distribution functions in the parallel direction, and $\langle v_{\perp e}^2\rangle/2$ and $\langle v_{\perp i}^2\rangle/2$ are the electron and ion thermal velocities in the perpendicular direction. In Eqs. (2.49) and (2.50), if we take a limit of zero temperature by putting $\langle v_\perp^2\rangle = 0$ and $f_{0||} = \delta(v_{||})$, we can recover the familiar dispersion relations for electron-cyclotron and ion-cyclotron waves, respectively.

We can see that the waves represented by Eqs. (2.49) and (2.50) are independent of the electrostatic waves with $k_\perp = 0$ discussed in 2.2a, because here only the transverse components of the electric field are involved, while previously, only the longitudinal component was involved. Therefore, we deduce that if a wave propagates along a field line, the transverse (electromagnetic) wave decouples from the longitudinal (electrostatic) wave.

Although we will discuss only the two stream type instability here, the dispersion relation obtained above is quite general and will be used later to discuss instabilities associated with an anisotropic velocity distribution. To look at a two stream type instability it is simplest to consider the case of a cold stream. As in the case of the electrostatic instability, there are basically two kinds of two streaming, one is the two humped distribution in electron (or ion) velocity distribution and the other is the existence of a relative average speed between electron and ion species. Unlike the electrostatic case, there exists *no* electromagnetic instability due to two streams in one species as will be shown later. Hence we consider here the latter case. For this we choose a coordinate system fixed to the ions such that the distribution functions are given by

$$\begin{aligned} f_{0\|}^{(e)} &= \delta(v_\| - v_0), \\ f_{0\|}^{(i)} &= \delta(v_\|). \end{aligned} \tag{2.51}$$

Substituting Eq. (2.51) into (2.50) we obtain the dispersion relation of the ion-cyclotron waves with drifting cold electrons

$$k^2 c^2 - \omega^2 + \omega_{pe}^2 \frac{\omega - k v_0}{\omega - k v_0 + \omega_{ce}} + \omega_{pi}^2 \frac{\omega}{\omega - \omega_{ci}} = 0. \tag{2.52}$$

The above expression shows that the dispersion relation for an electromagnetic wave in a vacuum, $k^2 c^2 - \omega^2 = 0$, is modified by the presence of the plasma. In the absence of ions ($\omega_{pi} = 0$), Eq. (2.52) gives the Doppler shifted backward propagating ($\omega < 0$ if $v_0 = 0$) whistler wave, which is called the slower cyclotron wave. Because this wave has a phase velocity slower than the drift speed, it carries negative energy. On the other hand, in the absence of the electron stream, Eq. (2.52) gives the dispersion relation for the ion cyclotron wave having resonance at $\omega \sim \omega_{ci}$, and this is the positive energy wave. The instability occurs by the coupling between the slower electron cyclotron wave and the ion cyclotron wave. The instability condition is obtained using Eq. (2.52) and solving for complex ω.

If the plasma density is reasonably large such that $\omega_{pi}/\omega_{ci} \gg 1$ (or equivalently $v_A (= c\omega_{ci}/\omega_{pi}) \ll c$, v_A is the Alfvén speed), the second term is negligible and the resultant equation becomes quadratic in ω and reduces to

$$\left(\frac{\omega}{\omega_{ci}}\right)^2 + \frac{\omega}{\omega_{ci}}\left(\frac{k^2 v_A^2}{\omega_{ci}^2} - \frac{k v_0}{\omega_{ci}}\right) - \left(\frac{k^2 v_A^2}{\omega_{ci}^2} - \frac{k v_0}{\omega_{ci}}\right) = 0. \tag{2.53}$$

It is then easily seen that ω becomes imaginary when $v_0 > k v_A^2/\omega_{ci}$. However this result is superficial because we have ignored the effect of a *dc* magnetic field generated by the current in the plasma. The current will generate a transverse magnetic field and the resultant total field becomes helical. This will violate our assumption of parallel propagation. To consider wave propagation in a nonuniform and helical *dc* magnetic field is a rather difficult task. However, this may lead to an instability of the current caused by the helically wound magnetic field line called the Kruskal-Shafranov instability (cf. Section 3.5), the instability condition of which is given by:

$$\frac{B_\theta}{B_0} > \frac{2\pi a}{L}. \tag{2.54}$$

If we use the identities $e n v_0 \pi a^2 = I_0$, B_θ (azimuthal magnetic field) $= \mu_0 I_0/2\pi a$ (a is the cross sectional radius of the current) and $k = 2\pi/L$ (L is the length of the pinch), Eq. (2.54) becomes

$$v_0 > 2 k v_A^2/\omega_{ci}. \tag{2.55}$$

Eq. (2.55) shows a drift speed twice as large as that obtained above assuming a straight magnetic field. Although the Kruskal-Shafranov instability is derived assuming an infinitely thin surface current, and hence is quite a different model from what we are treating here, the similarity of the result indicates that the same instability may be involved.

On the other hand if the total current is compensated for, as when a charge neutral plasma is moving with respect to another charge neutral plasma, the instability naturally becomes different from the current driven case above. The dispersion relation is obtained by adding to Eq. (2.52) additional terms corresponding to moving ions and stationary electrons. Then the instability condition is approximately given by:

$$v_0 > 2 v_A \left(\frac{n_0}{n_0 + n_s}\right)^{1/2} \tag{2.56}$$

(Hasegawa 1972), where n_s is the stream density and v_A is the Alfvén speed in the plasma. Note that in this case there exists a finite threshold drift speed, while in the previous example (Eq. (2.55)), the threshold speed can be zero for $k \to 0$. Therefore unlike the electrostatic case, the electromagnetic electron-ion two stream instability has a distinctive difference in its nature between the current compensated case and the uncompensated case.

When a weak stream of electrons (or ions) is intermixed with a background plasma, we may ignore the effect of the stream-generated magnetic field and the instability appears as a result of the coupling between the slower electron (ion) cyclotron mode in the stream and the ion (electron) cyclotron mode in the plasma.

Such an instability was first discussed by Bernstein and Dawson (1958) for $\omega \sim \omega_c$. However, Briggs (1964) as well as Hasegawa and Birdsall (1964) later pointed out that there are two ranges in ω ($\omega \approx 0$ and $\omega \approx \omega_{ci}$) and k ($k \dot{\approx} 0$ and $k \approx \omega_{ce}/v_0$) in which such an instability arises. This can be seen in Fig. 9, where we plot the relevant dispersion relation with an electron stream contribution added;

$$k^2 c^2 - \omega^2 + \omega_{pe}^2 \frac{\omega}{\omega + \omega_{ce}} + \omega_{pi}^2 \frac{\omega}{\omega - \omega_{ci}} + \omega_{ps}^2 \frac{\omega - k v_0}{\omega - k v_0 + \omega_{ce}} = 0 . \tag{2.57}$$

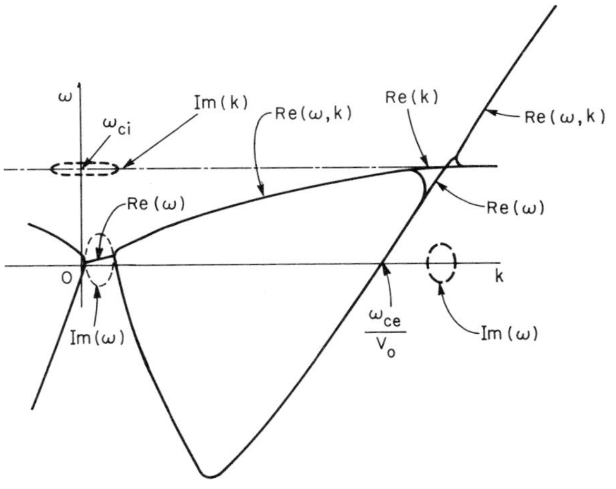

Fig. 9. Dispersion diagram for coupling between the ion-cyclotron wave and the slow electron-cyclotron wave in a two-stream electron-ion system

As shown in this figure, there are two different frequency ranges in which the instabilities are excited; we discuss them separately. The first type is the one excited near the cyclotron frequency of the stationary species. We call this the electromagnetic cyclotron wave instability. When this instability is applied to the magnetosphere, it can explain emissions of electromagnetic waves at kHz range (excitation of electron cyclotron wave by a proton beam (Kimura, 1961) or by an electron beam with a large perpendicular kinetic energy (Hruska, 1966; Liemohn, 1968)) or at a few Hz range (excitation of ion cyclotron wave by an electron beam (Jacobs and Higuchi, 1969; Criswell, 1969)).

On the other hand a simple observation of whistler noise does not prove the excitation by a proton beam, because as will be shown in the next section the whistler wave becomes unstable when the perpendicular kinetic energy is much larger than the parallel kinetic energy. It requires the simultaneous measurement of streaming particles. In this respect, the observation of ELF noise associated with auroral electron precipitation by Gurnett and Frank (1972) is a good indication of the electromagnetic electron-proton two stream instability.

Eviatar and Wolf (1968) have used this instability to explain turbulent viscous interaction between the solar wind and the magnetosphere. As will be discussed in Section 3.4a in relation to the Kelvin-Helmholtz instability, the interface between the solar wind and the magnetosphere in the morning and evening side does not have a dynamic equilibrium (Parker, 1967). Eviatar and Wolf proposed that the electromagnetic two stream cyclotron instability would produce electromagnetic turbulence and a consequent dissipation layer to maintain the equilibrium.

We now discuss the second type of the instability, an instability that occurs at frequencies much smaller than the cyclotron frequency (complex ω solution near the origin of the $\omega-k$ diagram in Fig. 9). We call this the Alfvén wave instability. We consider a situation in which an electron stream with a density n_s is mixed with a background plasma. If we take a long wavelength $k \ll \omega_{ce}/v_0$ and low frequency ($\omega \ll \omega_{ci}$) limit in the dispersion relation (2.57) we obtain:

$$k^2 - \frac{\omega^2}{v_A^2} + \frac{\omega_{ps}^2}{c^2 \omega_{ce}} (\omega - k v_0) = 0, \qquad (2.58\text{a})$$

or alternatively,

$$\left(\frac{\omega}{\omega_{ci}} - \frac{\alpha}{2}\right)^2 - \left(\frac{k v_A}{\omega_{ci}} - \frac{\alpha v_0}{2 v_A}\right)^2 = \frac{\alpha^2}{4}\left(1 - \frac{v_0^2}{v_A^2}\right) \qquad (2.58\text{b})$$

where $\alpha = n_s/n_0$ and $v_A (= c\,\omega_{ci}/\omega_{pi})$ is the Alfvén speed of the stationary plasma. It is readily seen that the instability occurs for $k = \alpha v_0 \omega_{ci}/2 v_A^2$

and the threshold condition is $v_0 > v_A$. The interesting point of this instability is that the frequency at which the instability occurs is given by $\alpha\omega_{ci}/2$, that is, it is proportional to the density ratio of the stream to the plasma. This fact was used by Nishida (1964) to explain irregular magnetic pulsations excited by precipitating auroral electrons and by Kimura and Matsumoto (1968) to explain pc 5 magnetic pulsations.

The unique perturbations of the magnetic field in the solar wind observed by Russell *et al.* (1971) shown in Fig. 10 may also be explained by the two stream electromagnetic instability of a compensated current case (Hasegawa, 1972).

We have considered so far mostly cases of parallel propagation. When we allow the wave to propagate obliquely to the background magnetic field, electromagnetic Bernstein waves are excited. The excited wave is mainly electrostatic in nature if $\beta \lesssim (m_e/m_i)^{1/2}$, as shown in the previous subsection, but becomes electromagnetic for a higher value of β. When we consider a case of completely perpendicular propagation ($k_{||}=0$), a new instability arises for a wave with an electric field polarized parallel to the magnetic field (ordinary mode). This instability (Momata, 1966) however is better classified as one generated by anisotropic velocity distributions (Weibel instability, (Weibel, 1959)) because it is not the directed velocity but the kinetic energy spread in the direction parallel to the magnetic field that causes the instability.

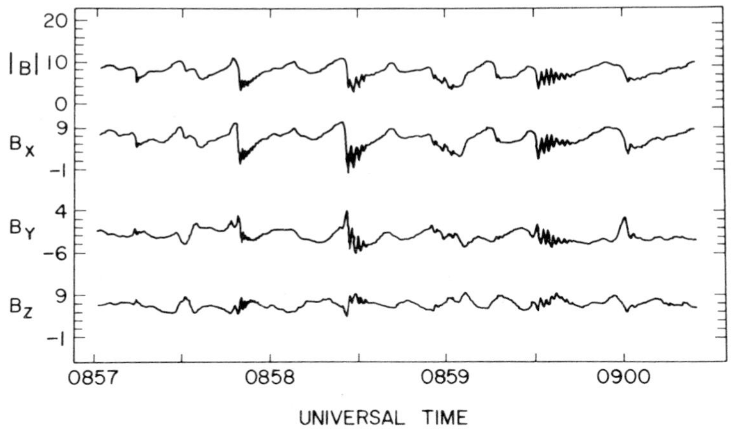

Fig. 10. OGO-5 satellite observation of discrete wave packets occurring in association with irregular low frequency waves on March 10, 1968, while the satellite was at 19.1 R_E and at a sun-earth-satellite angle of 67.5°. The X, Y and Z components are in the spacecraft coordinate system (Russell *et al.*, 1971)

2.2 c Effect of Magnetic Mirror

We have assumed so far that the background magnetic field is uniform and straight. However, if the wavelength of the excited perturbation is comparable to the typical scale length of the spatial variations of the background magnetic field, we must take into account effects arising from such a nonuniformity. There are basically two important effects. One is the drift of particles due to a gradient of the flux density or a curvature of the field line. Like in the case of Landau damping, this causes a phase mixing of a wave having a component of wave number perpendicular to the magnetic field, $k_\perp$, and a frequency comparable to $k_\perp v_D$, (v_D is the average drift speed) because such a drift speed is energy dependent. This effect will be discussed later in relation to the drift wave instability. The other effect is the bounce oscillation of particles trapped in the magnetic mirror. When a beam of particles with an average parallel velocity v_0 is trapped by the mirror and when the bounce frequency $\omega_b \sim v_0/z_0$ (z_0 is the mirroring distance from the center of the mirror) becomes comparable to the frequency of the perturbation, this effect is particularly important.

In this subsection we discuss an instability generated by such a bouncing beam of electrons (Nishihara *et al.*, 1972). For simplicity we consider an electrostatic perturbation propagating parallel to the magnetic field line. The Vlasov equation that describes the distribution function of particles trapped in the magnetic mirror can then be written approximately in one dimensional form, with a magnetic flux density changing as a function of coordinate z along the local field line:

$$\frac{\partial f}{\partial t}+v_{\|}\frac{\partial f}{\partial z}-\frac{e}{m_e}(\boldsymbol{E}+\boldsymbol{v}\times\boldsymbol{B})\cdot\frac{\partial f}{\partial \boldsymbol{v}}=0. \tag{2.59}$$

The unperturbed distribution function f_0 is obtained by putting $\partial f_0/\partial t=0$ as a function of constants of motion, the energy w and the magnetic moment μ,

$$f_0=f_0(w, \mu) \tag{2.60}$$

where

$$w=\frac{m_e}{2}(v_{\|}^2+v_\perp^2) \tag{2.61}$$

and

$$\mu=\frac{m_e v_\perp^2}{2B_0(z)}. \tag{2.62}$$

The perturbed distribution function $f_1(\boldsymbol{v}, z, t)$ then satisfies the linearized Vlasov equation

$$\frac{df_1}{dt} = \frac{\partial f_1}{\partial t} + v_{\|} \frac{\partial f_1}{\partial z} - \frac{e}{m_e} (\boldsymbol{v} \times \boldsymbol{B}_0) \cdot \frac{\partial f_1}{\partial \boldsymbol{v}} = \frac{e}{m_e} E_1 \frac{\partial f_0}{\partial v_{\|}}. \tag{2.63}$$

As before, f_1 can be obtained by integrating the right hand side along the unperturbed trajectory. Taking into account the fact that $\partial f_0/\partial w$ is constant, we have

$$\begin{aligned} f_1(\boldsymbol{v}, z)\, \mathrm{e}^{-i\omega t} &= \frac{e}{m_e} \int_{-\infty}^{t} dt' E(z') \mathrm{e}^{-i\omega t'} \left. \frac{\partial f_0}{\partial v_{\|}} \right|_{v_{\|} = v'_{\|}} \\ &= e \frac{\partial f_0}{\partial w} \int_{-\infty}^{t} dt' E(z') v'_{\|} \mathrm{e}^{-i\omega t'}, \end{aligned} \tag{2.64}$$

and the unperturbed trajectory is specified by the equation of motion

$$\ddot{z}' = -\frac{\mu}{m_e} \frac{\partial B_0(z')}{\partial z'}. \tag{2.65}$$

For simplicity we assume the beam to be trapped near the central (equatorial if in the magnetosphere) region of the mirror where $B_0(z)$ may be taken to be parabolic:

$$B_0(z) = B_0(1 + b z^2). \tag{2.66}$$

Then Eq. (2.65) can be integrated immediately to give

$$z' = z_0 \sin(\omega_b t' + \varphi) \tag{2.67}$$

and

$$v'_{\|} = z_0 \omega_b \cos(\omega_b t' + \varphi) \tag{2.68}$$

where the bounce frequency ω_b is given by

$$\omega_b = \left(\frac{2 \mu b B_0}{m_e} \right)^{1/2}. \tag{2.69}$$

The mirror point z_0 and the phase φ must be decided by the boundary (final) condition at $t'=t$, at which time $z'=z$ and $v'_{\|}=v_{\|}$. This gives

$$z_0=\left(z^2+\frac{v_{\|}^2}{\omega_b^2}\right)^{1/2}=\frac{1}{\sqrt{b}}\left(\frac{w}{\mu B_0}-1\right)^{1/2} \tag{2.70}$$

and

$$\omega_b t+\varphi=\pm\sin^{-1}\frac{z}{z_0}. \tag{2.71}$$

$\pm$ signs in the above expression corresponds to $v_{\|}\gtrless 0$ respectively and distinguishes particles going left and right along the z axis.

Because of the nonuniformity in the z direction, the z dependency of the electric field $E(z)$ is not a pure sinusoidal ($\sim e^{ikz}$) function as were the cases treated in the previous subsections. Hence we express $E(z)$ here in terms of a Fourier series with a fundamental period of length $2L$, L may be taken to be the distance z at which most of the beam particles are reflected, i.e.,

$$E(z)=\sum_{n=-\infty}^{\infty} E_n e^{inkz}, \tag{2.72}$$

$$E_n=\frac{1}{2L}\int_{-L}^{L} E(z)\,e^{-inkz}dz \tag{2.73}$$

and

$$k=\frac{\pi}{L}.$$

If we use the above expressions for $E(z)$ in Eq. (2.64) together with the z' and $v'_{\|}$ from Eqs. (2.67) and (2.68), we can integrate with respect to t' to give

$$f_1^{\pm}(\boldsymbol{v},z)=-ie\frac{\partial f_0}{\partial w}\sum_{n=-\infty}^{\infty} E_n \sum_{l\neq 0}^{\pm\infty}\frac{J_l(nkz_0)\exp\left(il\sin^{-1}z/z_0\right)}{nk\left(1\mp\dfrac{\omega}{l\omega_b}\right)}, \tag{2.74}$$

where $f_1^{\pm}$ represents the perturbed distribution function for $v_{\|}\gtrless 0$ respectively. Note here that z_0 is a function of w and μ, while ω_b is a function of μ only. Although the distribution functions obtained above have different expressions for $v_{\|}\gtrless 0$, they can be seen to be continuous at $v_{\|}=0$, where $z=z_0$ and $\exp\left(il\sin^{-1}z/z_0\right)=i^l$.

The perturbed charge density $\varrho_1(z)\mathrm{e}^{-i\omega t}$ can be obtained from Eq. (2.74) by integrating over velocity space. Because f_1 is an even function of $v_{||}$, $\varrho_1(z)$ becomes

$$\begin{aligned}\varrho_1(z) &= -en_0 \int_{-\infty}^{0} dv_{||} \int_0^{\infty} 2\pi v_\perp dv_\perp f_1^- - en_0 \int_0^{\infty} dv_{||} \int_0^{\infty} 2\pi v_\perp dv_\perp f_1^+ \\ &= -en_0 \int_0^{\infty} dv_{||} \int_0^{\infty} 2\pi v_\perp dv_\perp (f_1^+ + f_1^-).\end{aligned} \tag{2.75}$$

Since f_0 is a function of w and μ it is convenient here to convert the volume element $dv_\perp dv_{||}$ into $d\mu\, dw$ through

$$2\pi v_\perp dv_\perp dv_{||} \propto 2\pi \frac{B_0(z)}{m_e^2 v_{||}} d\mu dw = \pi B_0(z) \left[\frac{2}{m_e^3 (w - \mu B_0(z))}\right]^{1/2} \mathrm{d}\mu dw. \tag{2.76}$$

The normalization of f_0 is

$$\int_0^{\infty} d\mu \int_{\mu B_0(z)}^{\infty} dw f_0(w, \mu) \frac{1 + bz^2}{[w - \mu B_0(z)]^{1/2}} = 1 \tag{2.77}$$

at $z = 0$.

Eq. (2.75) then becomes

$$\begin{aligned}\varrho_1(z) = 2i \sum_{\mathrm{n}} \frac{e^2 n_0}{\mathrm{n}k} \int_0^{\infty} d\mu \int_{\mu B_0(z)}^{\infty} dw \frac{\partial f_0}{\partial w} \frac{1 + bz^2}{[w - \mu B_0(z)]^{1/2}} \\ E_{\mathrm{n}} \sum_{l \neq 0}^{\pm\infty} \frac{J_l(\mathrm{n}kz_0) \exp(il \sin^{-1} z/z_0)}{1 - \omega^2/l^2 \omega_b^2}.\end{aligned} \tag{2.78}$$

If we substitute the charge density into Poisson's equation

$$\nabla \cdot \boldsymbol{E} = \frac{\varrho_1}{\varepsilon_0}, \tag{2.79}$$

we can obtain the dispersion relation. $\nabla \cdot \boldsymbol{E}$ in the above expression must be evaluated by taking into account the converging field lines,

$$\nabla \cdot \boldsymbol{E} = \nabla \cdot \left[\frac{(\boldsymbol{E} \cdot \boldsymbol{B}_0)}{|\boldsymbol{B}_0|^2} \boldsymbol{B}_0\right] = \boldsymbol{B}_0 \cdot \nabla \frac{(\boldsymbol{E} \cdot \boldsymbol{B}_0)}{|\boldsymbol{B}_0|^2} = (1 + bz^2) \frac{\partial}{\partial z} \left[\frac{E(z)}{1 + bz^2}\right]. \tag{2.80}$$

When this expression is substituted into (2.79), the $(1+bz^2)$ factors in Eqs. (2.80) and (2.78) cancel. If we then Fourier expand the resultant expression of (2.79), the dispersion relation is finally obtained in the form of an infinitely coupled mode;

$$|\delta_{mn}+\varepsilon_{mn}|=0,$$

where

$$\delta_{mn}=0\,;\ n\neq m$$
$$\delta_{mn}=1\,;\ n=m \tag{2.81}$$

and

$$\varepsilon_{mn}=\frac{1}{1-bL^2/3}\sum_{\text{species}}\frac{m_e\omega_{p0}^2}{2k^2 mnL}\int_{-L}^{L}dz\,e^{-imkz}\int_0^\infty d\mu\int_{\mu B_0(z)}^\infty\frac{dw}{[w-\mu B_0(z)]^{1/2}}$$
$$\cdot\frac{\partial f_0}{\partial w}\sum_{l=-\infty}^{\infty}\frac{l^2\omega_b^2 J_l(nkz_0)\exp(il\sin^{-1}z/z_0)}{\omega^2-l^2\omega_b^2}. \tag{2.82}$$

In Eq. (2.82) $(1+bz^2)^{-1}$ was approximated by $(1-bz^2)$ in the process of the Fourier expansion and ω_{p0} is the plasma frequency at $z=0$.

If we compare the dielectric constant obtained from Eq. (2.82) with that of the uniform magnetic field, Eq. (2.4), we can see that a significant complication is caused by the nonuniform magnetic field. In particular we can see that a resonant wave – particle interaction exists at frequencies corresponding to the harmonics of the bounce frequency ω_b. Because ω_b depends on the magnetic moment, integration over μ in Eq. (2.82) produces a phase mixing similar to that in Landau damping. This fact is easily seen when we remember that $\omega_b\sim k_{\|}v_{\|}$. If the distribution function is monochromatic, such phase mixing is absent. To study the contributions of a trapped beam of electrons, we assume for the unperturbed distribution function $f_0(w,\mu)$ a monochromatic distribution both in energy and magnetic moment such as

$$f_0(w,\mu)=\delta(w-w_0)\,\delta(\mu-\mu_0)\,(w_0-\mu_0 B_0)^{1/2}. \tag{2.83}$$

The last factor in this expression is the normalization factor that originates from expression (2.77). Substituting Eq. (2.83) into the expression for the dielectric constant ε_{mn} (Eq. (2.82)) and performing the integration in the order of μ, z and w by suitably modifying the domain of integration, we obtain

$$\varepsilon_{\text{nn}} = \sum_{l=0}^{\infty} \frac{2l^2\omega_{p0}^2}{\text{n}(l^2\omega_{b0}^2 - \omega^2)} \frac{J_l(\lambda)J_l'(\lambda)}{1 - bL^2/3}. \tag{2.84a}$$

In this expression, we have shown only the diagonal term of ε_{mn} because the off diagonal terms are small for a weak nonuniformity. Also in the above expression,

$$\begin{aligned} \lambda &= \text{n}k z_0(\mu_0, w_0) \\ &= \text{n}k \frac{1}{\sqrt{b}} \left(\frac{w_0}{\mu_0 B_0} - 1\right)^{1/2} \\ &= \frac{\text{n}k v_{\|0}}{\omega_{b0}} \\ &= \frac{\text{n}\pi z_0}{L}, \end{aligned} \tag{2.85}$$

$$\omega_{b0} = \left(\frac{2\mu_0 b B_0}{m_e}\right)^{1/2} \tag{2.86a}$$

and $v_{\|0}$ is the parallel velocity of electron beam at $z=0$. The dielectric constant obtained above has a form similar to that of the Bernstein wave because of the similarity of the periodic bounce motion of a particle in the mirror field and cyclotron motion. Let us see the sign of energy of a wave in a medium with this dielectric constant. By taking the derivative with respect to ω, we can see that ε_{nn} in Eq. (2.84a) represents a negative energy medium if

$$J_l(\lambda)J_l'(\lambda) < 0. \tag{2.87a}$$

Because J_l is an oscillatory function of λ, there are repeated ranges of λ in which $J_l J_l'$ becomes negative. For example, for $l=1$, the negative energy wave appears for $\lambda = \text{n}k z_0$ that satisfies

$$\begin{aligned} &1.84 < \lambda < \ 3.83, \qquad && 5.33 < \lambda < \ 7.02, \\ &8.54 < \lambda < 10.17 && 11.71 < \lambda < 13.32 \ldots. \end{aligned} \tag{2.87b}$$

In other words, given a value of z_0 which is the distance from $z=0$ to the reflection point of a beam having the energy w_0 and magnetic moment μ_0, if we choose a wave number nk such that λ $(=nkz_0)$ falls within these intervals, the wave $\sim e^{inkz}$ carries negative energy.

For example, if we take $\lambda=2.8$ as a nominal value of λ that falls in the first interval in inequality (2.87b) giving a maximum negative value of $J_1 J_1'$ and take $n=1$ and ignore $bL^2/3$, we have $J_1 J_1' = -0.14$ and near $\omega \sim \omega_{b0}$ the dielectric constant (2.84a) becomes

$$\varepsilon_{11} \sim \frac{0.28\,\omega_{p0}^2}{\omega^2 - \omega_{b0}^2} \tag{2.84b}$$

with $k=2.8/z_0$. If such a beam alone is trapped in the field, the dispersion relation is given by

$$1 + \frac{0.28\,\omega_{p0}^2}{\omega^2 - \omega_{b0}^2} = 0 \tag{2.88a}$$

and the beam becomes unstable when $\omega_{p0} \gtrsim 2\,\omega_{b0}$. Because the bounce frequency is rather small (much smaller than the cyclotron frequency) the above condition is satisfied rather easily. This example is an analog of the so called counter stream instability (two stream instability with two streams with the same magnitude but an opposite velocity), in a uniform plasma, whose dispersion relation is given by

$$1 - \frac{0.5\,\omega_{p0}^2}{(\omega - k v_0)^2} - \frac{0.5\,\omega_{p0}^2}{(\omega + k v_0)^2} = 0. \tag{2.88b}$$

We can see how the "trapped" two stream modifies its dispersion relation by comparing Eq. (2.88a) with (2.88b).

In the magnetosphere, long period $(\omega \ll \omega_{ci})$ electrostatic oscillations (in the form of particle flux variations) are sometimes observed as shown in Fig. 11. With respect to frequency this oscillation falls in the pc 3 range but because there has been no simultaneous variation of the magnetic field (Nishihara *et al.*, 1972), it is difficult to identify this as an Alfvén wave oscillation (which couples an electrostatic mode through a finite β effect). It is also difficult to consider such an oscillation to be that of an ion acoustic wave because ordinarily in the magnetosphere the ion temperature (~ 10 keV) is larger than the electron temperature (~ 1 keV) and an ion acoustic wave is heavily damped.

Let us see, then, if a trapped beam of electrons can cause such an oscillation. Because of the resonant property of the dielectric constant of

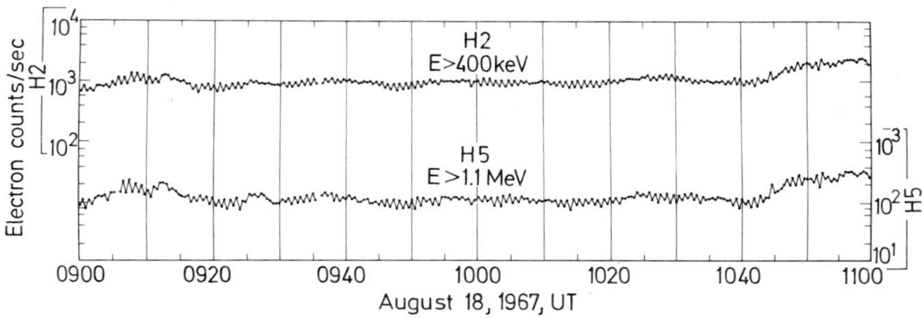

Fig. 11. Short period (~54 sec) fluctuations in the electron fluxes ($E>0.4$, >1.1 MeV) measured on ATS-1 during 09-11 UT (23-01 LT) August 18, 1967. Distinct beats between the instrument sampling time and the electron modulations are discernible (Nishihara *et al.*, 1972)

a trapped beam at $\omega = n\omega_{b0}$, the instability, if excited, is expected to produce oscillations at these resonant frequencies. Therefore, let us first evaluate the bounce frequency of an electron beam with perpendicular energy $w_{0\perp}$ trapped in the geomagnetic field. As can be seen from Eq. (2.69), we need only the perpendicular energy to calculate ω_b,

$$\omega_{b0}^2 = \frac{2\mu_0 b B_0}{m_e} = \frac{2 w_{0\perp} b}{m_e} = v_{0\perp}^2 b. \tag{2.86b}$$

Hence we only need to find b, which can be defined

$$b = \frac{1}{2B_0} \left. \frac{\partial^2 B(s)}{\partial s^2} \right|_{s=0}, \tag{2.89}$$

where s is the distance along the field line measured from the equator.

For a dipole field the equation of the field line is given by

$$r = R_0 \cos^2 \varphi \tag{2.90}$$

where φ is the complement of the polar angle and R_0 is the equatorial distance to the field line. The flux density is given also as a function of φ and r:

$$B = \frac{R_0^3}{r^3} B_0 (1 + 3\sin^2 \varphi)^{1/2}. \tag{2.91}$$

Using Eq. (2.90) the distance s along B can be obtained as

$$s=\int_0^{\varphi}\left[r^2+\left(\frac{dr}{d\varphi}\right)^2\right]^{1/2}d\varphi$$

$$=\int_0^{\varphi} R_0(1+3\sin^2\varphi)^{1/2}\,d(\sin\varphi), \tag{2.92}$$

$$\therefore ds=R_0(1+3\sin^2\varphi)^{1/2}\,d(\sin\varphi).$$

The value of b is then obtained using Eqs. (2.91) and (2.92),

$$b=\frac{1}{2B_0}\left.\frac{\partial^2 B}{\partial s^2}\right|_{\varphi=0}=\frac{9}{2}\frac{1}{R_0^2}. \tag{2.93}$$

The bounce frequency of the beam is therefore given by

$$\omega_{b0}^2=\frac{2\mu_0 B_0}{m_e}\frac{9}{2}\frac{1}{R_0^2}$$
$$=\frac{9\,w_{0\perp}}{m_e R_0^2}. \tag{2.94}$$

Since R_0 for the synchronous altitude is approximately 4×10^7 m, for a 54 sec period, the perpendicular energy of the trapped beam of electrons becomes ~ 13 eV. The parallel energy is somewhat arbitrary and can even be almost zero.

It is not difficult to show that such a beam of trapped electrons causes an instability at $\omega\sim\omega_{b0}$ when it is intermixed with the background plasma. For example if we assume that $v_{0\parallel}$ is sufficiently small so that $k v_{Ti}\gg\omega_{b0}$, the background dielectric constant ε_b may be given by

$$\varepsilon_b=\frac{\omega_{pe}^2}{k^2 v_{Te}^2}+\frac{\omega_{pi}^2}{k^2 v_{Ti}^2}+i\left(\frac{\pi}{2}\right)^{1/2}\frac{\omega_{pi}^2}{k^2 v_{Ti}^2}\frac{\omega}{k v_{Ti}}, \tag{2.95}$$

where usual notations are used for the background plasma. The unstable solution is obtained by combining Eqs. (2.95) and (2.84b) so that $\varepsilon_b+\varepsilon_{11}=0$ and

$$\omega\sim\omega_{b0}+i\frac{\omega_{p0}^2}{k v_{Ti}}\frac{T_e}{T_i+T_e}. \tag{2.96}$$

In this way we find a growing electrostatic wave with $\omega\sim\omega_{b0}$.

We have shown in this subsection that the magnetic mirror produces a significant modification to the dispersion relation obtained for a uniform field when the wave length is comparable to the scale length of the mirror. We have treated the effect only for a monochromatic beam of particles; however, effects for trapped plasma without a monochromatic energy are also important. In particular when we discuss waves with frequency lower than $\langle\omega_b\rangle$ and wave length comparable to the length of the field line (or its radius of curvature), the small argument expansion shown in Eq. (2.14) is inapplicable. Instead of the denominator $(\omega - kv)^{-1}$ we must use $(\omega^2 - \omega_b^2)^{-1}$ as shown in Eq. (2.82). The most important consequence of this change is that the imaginary part of ε, which was proportional to ω/kv_T in the case of the straight field line, vanishes for the lowest order of ω/ω_b. This means, for the lowest order of $\omega/\omega_b \sim \omega/k_{\|} v_T$, the Landau damping disappears for waves in the trapped particles!

Let us see how this can occur. The dielectric constant (susceptibility) ε of a plasma in a magnetic mirror for $\omega < \langle\omega_b\rangle$ can be written approximately from Eq. (2.82) as

$$\begin{aligned} \varepsilon &= \int_0^\infty \frac{f(\omega_b^2)}{\omega^2 - \omega_b^2}\, d\mu \propto \int_{-\infty}^{\infty} \frac{\omega_b f(\omega_b^2)}{\omega^2 - \omega_b^2}\, d\omega_b \\ &= \frac{1}{2\omega} \int_{-\infty}^{\infty} \left(\frac{1}{\omega_b - \omega} - \frac{1}{\omega_b + \omega} \right) \omega_b f(\omega_b^2)\, d\omega_b, \end{aligned} \tag{2.97}$$

where $f(\omega_b^2)$ is an even function of ω_b because z_0 is an even function of ω_b. The above integral must be evaluated by using the fact that Im $\omega > 0$

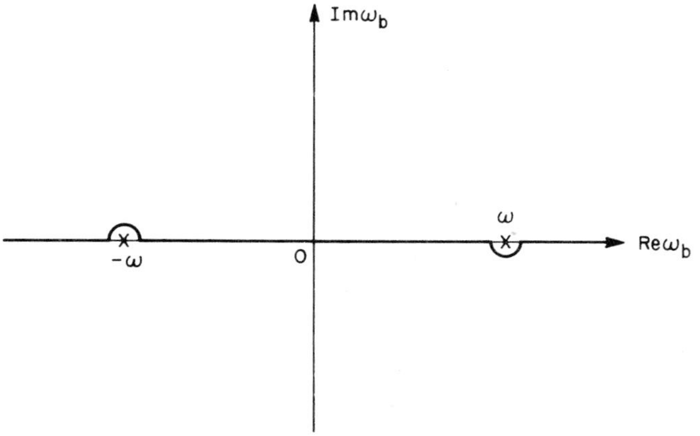

Fig. 12. Path of integration of Eq. (2.97) over variable ω_b

following the Laplace transformation in ω space. The poles are at $\omega_{\bar{b}}=\omega$ and $\omega_{\bar{b}}=-\omega$, and hence lie respectively above and below the real axis of $\omega_{\bar{b}}$. Consequently the integral path along the real axis must be deformed as shown in Fig. 12 when Im $\omega\to 0$. The imaginary part of ε can then be evaluated using the Dirac expression in a similar way to that used for Eq. (2.14)

$$\mathrm{Im}\,\varepsilon\sim\frac{1}{2\omega}\left[i\pi\omega f(\omega^2)+i\pi(-\omega)f(\omega^2)\right]=0\,.$$

Therefore in the lowest order of $\omega/\langle\omega_{\bar{b}}\rangle$, the imaginary part vanishes. This can be interpreted as a consequence of emission of waves once Landau damped by the periodic motion of the bouncing particles in the mirror field (Berk and Book, 1969). The absence of the Landau damping for the waves with $\omega<k_{\|}v_T\sim\omega_{\bar{b}}$ in the mirror field can be very important; for example, in the drift wave instability (Section 3.1) where the driving force is the dissipative effect of electrons due to the wave particle interaction. The above proof shows that the drift wave instability in the mirror field cannot exist, at least in the lowest order of $\omega/\omega_{\bar{b}}$, in its ordinary form with excitation by wave particle interaction (electron Landau damping).

2.2d Application to the Field Aligned Current – Anomalous Resistivity

We have seen that when the plasma electrons drift with respect to ions with an average speed larger than some threshold value, both the electrostatic and electromagnetic field grow at the expense of the electron kinetic energy. When the instability occurs, the field energy grows exponentially; therefore the loss of kinetic energy of electrons is also exponential and the current carried by these electrons is suddenly disrupted. Such a phenomenon has been noticed in laboratory plasma discharge experiments (for example, Hamberger and Jancarik, 1970) and is still one of the most controversial subjects in recent research in high temperature plasmas.

The existence of a large field aligned current during auroral substorm activities (Cloutier *et al.*, 1970; Cummings and Dessler, 1967; Zmuda *et al.*, 1966) has led to the speculation that this current may generate an instability which may cause an anomalously large resistivity which could create a very large electric field parallel to the magnetic field. Akasofu (1969) postulated that this electric field would produce acceleration of electrons and lead to the aurora breakup. Swift (1965) considered the possibility that an ion acoustic wave instability might produce such an effect, while Kindel and Kennel (1971) proposed the application of electrostatic ion cyclotron instability. Observations of ELF noise in the auroral region (Gurnett and Frank, 1972; F. L. Scarf *et al.*, 1973) seem to indicate the existence of a two stream instability. In particular the simultaneous observations of electrostatic emission and a field-aligned current by Scarf *et al.* as shown in Fig. 13 are good evidence of such an instability.

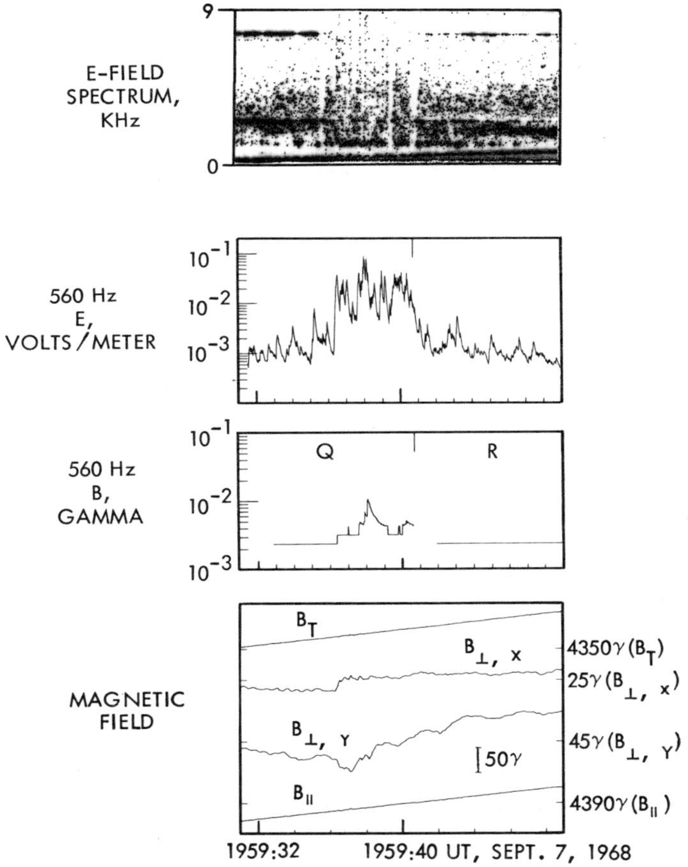

Fig. 13. Field and wave characteristics measured near the initial steep gradient in $E > 50$ keV electron flux by OGO-5. The magnetometer shows the instantaneous total field (B_T) and the orthogonal components in a frame of reference defined with respect to the average local magnetic field. The field aligned current system centered near 19h 59m 37s is associated with the detection of an intense low frequency (e.g., 560 Hz) electrostatic hiss band and with the detection of sporadic rising impulse noise bursts extending up to 9 kHz. (After Scarf *et al.*, 1973)

There are several other positions in the magnetosphere, in the solar corona or even in interplanetary space, where such a current generated instability might be excited. Such instabilities can produce "anomalous resistivity" and hence the assumption of infinite plasma conductivity breaks down. This presents a serious difficulty in applying MHD equations, which assume no electric field along the magnetic field.

In this subsection, let us consider how such anomalous resistivity is generated. It is easily shown that the electrical resistivity η can be obtained in terms of the electron-ion collision frequency ν_{ei} as

$$\eta = \frac{\nu_{ei}}{\varepsilon_0 \omega_{pe}^2} \, (\Omega \mathrm{m}), \tag{1.28}$$

where, as shown in Section 1, ν_{ei} is approximately given by

$$\nu_{ei} = \omega_{pe} \frac{1}{n_0 \lambda_{De}^3} = \omega_{pe} \frac{W_T}{n_0 T_e} \tag{2.98}$$

where W_T is the energy density of the fluctuating electrostatic field in *thermal equilibrium*. If the resistivity is given by Eq. (1.28), it can be assumed to be essentially zero in the magnetosphere. However, when the electron drift speed is larger than its thermal speed, the electron-ion two stream instability occurs. Buneman has considered that in such a case the electron-ion collision frequency is effectively increased to the growth rate of the instability, $\omega_{pe}(m_e/m_i)^{1/3}$ (Buneman, 1958) i.e.,

$$\nu_{\mathrm{eff}}^{(1)} = \omega_{pe} \left(\frac{m_e}{m_i} \right)^{1/3}. \tag{2.20}$$

Because ordinarily $(n\lambda_D^3)^{-1}$ is $\sim 10^{-3}$ to 10^{-4} while $(m_e/m_i)^{1/3} \sim 10^{-1}$, we can see a tremendous increase in the effective resistivity. However, because the rate of momentum loss is so large, the beam loses its kinetic energy in a few tens of electron plasma oscillation periods; simultaneously the electron distribution function spreads and the drift speed v_0 becomes smaller than the thermal speed v_{Te}. When v_0 is smaller than v_{Te} but still larger than the ion sound speed, $v_{Te}(m_e/m_i)^{1/2}$, the ion acoustic instability or the ion cyclotron instability will be generated. These instabilities have much smaller growth rate than the Buneman instability and hence may persist for a much longer period of time. As was shown, these instabilities are generated by a small number of resonant electrons. Hence the growth rate now cannot be used as the effective net loss rate of electron momentum, nor as the effective collision frequency. In such a case, we can obtain the momentum loss rate of the average electrons from the net momentum balance between electrons and waves.

As is known from the quantum theory of waves, the momentum of a wave quantum is given by $\hbar \boldsymbol{k}$. Hence when the growth rate is much smaller than the frequency of the wave one can introduce the concept of wave momentum density $\boldsymbol{P}_W$ defined by

$$\boldsymbol{P}_W = \sum_{\boldsymbol{k}} \hbar \boldsymbol{k} N_{\boldsymbol{k}} \tag{2.99}$$

where $\hbar$ is Planck's constant, $\boldsymbol{k}$ is the wave number of the wave and $N_{\boldsymbol{k}}$ is the number density of quasiparticles (e.g., plasmons, phonons, etc.) associated with the waves, defined by

$$N_{\boldsymbol{k}} = \frac{W_{\boldsymbol{k}}}{\hbar\omega} = \frac{1}{\hbar\omega} \frac{\varepsilon_0 |E_{\boldsymbol{k}}|^2}{2} \omega \frac{\partial \varepsilon}{\partial \omega} \tag{2.100}$$

where $W_{\boldsymbol{k}}$ is the energy density of the wave and $\hbar\omega$ is the energy of the wave quantum. When an instability occurs, the wave energy increases at the rate of $2\gamma_{\boldsymbol{k}}$ (where $\gamma_{\boldsymbol{k}}$ is the growth rate corresponding to the wave number $\boldsymbol{k}$) and hence the wave gains momentum by $2\gamma_{\boldsymbol{k}} \boldsymbol{P}_W$ per second. This rate of increase of the wave momentum density is provided by particles. Hence, the effective loss rate of particle momentum $\nu_{\text{eff}}^{(2)}$, which is the effective collision frequency, can be obtained by

$$\nu_{\text{eff}}^{(2)} m_e n_0 v_0 = \sum_k 2\gamma_k \hbar k N_k$$

or

$$\nu_{\text{eff}}^{(2)} = \sum_k \frac{2\gamma_k}{m n_0 v_0} \left(\frac{k}{\omega}\right) \frac{\varepsilon_0 |E_k|^2}{2} \omega \frac{\partial \varepsilon}{\partial \omega}. \tag{2.101}$$

Here we assumed for simplicity that the generated wave has a $\boldsymbol{k}$ spectrum concentrated mainly in the direction of v_0. However, the results still hold for $\boldsymbol{k}$ spectrum spreaded in a wide angle if we take a projection of the wave momentum in the direction of $\boldsymbol{v}_0$.

Let us now derive the above expression directly from the Vlasov equation. By averaging over coordinate space, the nonlinear change in the distribution function due to the instability can be obtained from the Vlasov equation as (cf. Section 4.1)

$$\frac{\partial \langle f \rangle}{\partial t} = \left\langle \frac{e}{m_e} \sum_k E_k \frac{\partial f_k}{\partial v} \right\rangle, \tag{2.102}$$

where the excited electric field is written as

$$E_1(x, t) = \sum_k E_k e^{i(kx - \omega t)}.$$

If we multiply Eq. (2.102) by v and integrate over velocity space, we obtain the time rate change of the average speed of electrons as

$$m n_0 \frac{\partial \langle v \rangle}{\partial t} \equiv -\nu_{\text{eff}}^{(2)} m_e n_0 v_0$$

$$= \sum_k \left\langle e n_0 E_k \int v \frac{\partial f_k}{\partial v} dv \right\rangle$$

$$= -\operatorname{Re} \sum_k e n_k E_k^*. \tag{2.103}$$

The perturbed charge density, en_k, is related to the perturbed electric field E_k and the dielectric constant $\varepsilon(\omega, k)$, by

$$e n_k = i k \varepsilon_0 \varepsilon E_k. \tag{2.104}$$

Substituting this expression into Eq. (2.103), we have

$$\nu_{\text{eff}}^{(2)} = -\sum_k \frac{2\varepsilon_0 k}{m_e n_0 v_0} \frac{|E_k|^2}{2} \operatorname{Im} \varepsilon. \tag{2.105}$$

As shown in Section 1 the growth rate γ_k is related to $\operatorname{Im}\varepsilon$: $\gamma_k = -\operatorname{Im}\varepsilon/(\partial\varepsilon/\partial\omega)$. If we use this growth rate in Eq. (2.105), we have Eq. (2.101). Therefore the effective collision frequency obtained by the physical argument is derived exactly using the Vlasov equation. (We note, however, that the use of $\nu_{\text{eff}}^{(2)}$ in Eq. (2.98) to calculate the effective dc resistivity may not be satisfactory because the quasilinear scattering occurs only for those electrons that satisfy the resonant condition $\boldsymbol{k} \cdot \boldsymbol{v} = \omega_s$, thus applies only to a minor portion of the electron distribution. To account for this difficulty, one should consider non-resonant scattering also. This is particularly important when the plasma is stable.)

As shown in Subsection 2.1a the growth rate of the ion acoustic wave γ_k can be approximated as

$$\gamma_k \sim \sqrt{\frac{m_e}{m_i}}\, \omega. \tag{2.18}^\star$$

⋆ This is for $\boldsymbol{k} \| \boldsymbol{v}_0$; for a wave vector oblique to $\boldsymbol{v}_0$, the growth rate has $\boldsymbol{k}$ dependency.

If we substitute this growth rate into Eq. (2.101) the effective collision frequency is expressed as

$$\nu_{\mathrm{eff}}^{(2)} = \sum_k \frac{2W_k}{n_0 T_e} \frac{k v_{Te}^2}{v_0} \sqrt{\frac{m_e}{m_i}}. \tag{2.106}$$

Now, for the excitation of the ion acoustic wave instability, $v_0 > c_s$. Thus if we use the critical velocity $v_0 = c_s$ and also $v_{Te}(m_e/m_i)^{1/2} = c_s$, and $k \sim \omega_{pe}/v_{Te}$, the above expression reduces to

$$\nu_{\mathrm{eff}}^{(2)} \sim \omega_{pe} \sum_k \frac{W_k}{n_0 T_e}. \tag{2.107}$$

If we compare the above expression with that for the classical electron-ion collision frequency in Eq. (2.98), we notice an extremely good resemblance. The effective collision frequency is given by the same expression as the classical case when the fluctuation field energy density in thermal equilibrium W_T is replaced by the field energy density generated by the instability $W = \Sigma W_k$.

Naturally the fluctuating field energy is minimal at thermal equilibrium, hence the effective collision frequency given above is always larger than the Coulomb collision frequency. Although the Eq. (2.107) is obtained for the particular case of the ion acoustic instability, the expression can be shown to be generally valid for a weak instability. The general proof can be made by considering a quasi-equilibrium situation in which the loss rate of the particle momentum ν_{eff} and the diffusion rate D of the particles in the velocity space (quasilinear diffusion caused by the generated wave) satisfy the relation $\nu_{\mathrm{eff}} = D/v^2$ (Tsytovich, 1970). This situation indicates that anomalous resistivity is a process by which a systematic (collective) transfer of the *dc* current energy to high frequency turbulent field energy occurs. This high frequency field energy in turn produces diffusion of particles in velocity space in a quasi-equilibrium manner. Note that, for this process to work as a resistivity, the wave momentum should eventually be absorbed by ions.

For the transfer of *dc* current energy to high frequency field energy, one can consider an alternative process to the two stream instability. In particular, as in the case of the field aligned current, when the current source has a large inertia (inductance), a sudden disruption of the current by a Buneman type instability can cause a transient *ac* electric field. This *ac* field attempts to compensate for the loss of the convection current by generating a displacement current. A decay instability (non-linear wave-wave interaction) of an *ac* electric field of this type can be shown (Nishihara and Hasegawa, 1972) to produce ion acoustic wave turbulence which can also cause anomalous resistivity.

Finally we must realize that anomalous resistivity is not the only consequence of the field aligned current. In fact the two stream instability has such a short wavelength that it cannot control a macroscopic form of the current pattern. In this respect, it is also important to consider large scale MHD instabilities. The Kruskal-Shafranov instability for example will produce a kink in the current path causing a large scale deformation of the current path.

2.3 Instabilities Due to Anisotropic Velocity Distributions

In the presence of a magnetic field in which the particle collision frequency is much smaller than the cyclotron frequency, the velocity distribution can be anisotropic with respect to the direction of the magnetic field. Anisotropic distributions can be categorized into two types. One type has perpendicular (with respect to the direction of the magnetic field) kinetic energy larger than parallel kinetic energy, while the other has parallel kinetic energy larger than perpendicular kinetic energy. For both cases instabilities can occur.

When the parallel kinetic energy is larger, an electromagnetic instability is generated. This instability can be related to the fire hose instability for $\omega \ll \omega_{ci}$, and will be treated in Section 2.4. In this section we will consider instabilities generated when the perpendicular kinetic energy is larger. In this case both electrostatic and electromagnetic perturbations can be excited. We discuss these in Subsections 2.3a and 2.3b; in Subsection 2.3c we discuss the combined effect of a density gradient and an anisotropic distribution and in Subsection 2.3d the consequences of these instabilities in the particle distribution function will be discussed.

2.3a Electrostatic Instabilities

We have derived the scalar conductivity for electrostatic perturbations in Subsection 2.2a:

$$\sigma(\omega, k) = \frac{i\omega\varepsilon_0\omega_p^2}{k^2} \sum_{n=-\infty}^{\infty} \int d\boldsymbol{v} \left[J_n\left(\frac{k_\perp v_\perp}{\omega_c}\right)\right]^2 \cdot \frac{k_{\|}(\partial f_0/\partial v_{\|}) + (n\omega_c/v_\perp)(\partial f_0/\partial v_\perp)}{k_{\|}v_{\|} - (\omega - n\omega_c)}. \tag{2.28}$$

Because the necessary condition of the instability is that the real part of the conductivity for real ω be negative, let us see under what circumstances the real part of Eq. (2.28) becomes negative. The real part appears because of the pole at $v_{\|}=(\omega-\mathrm{n}\,\omega_c)/k_{\|}$, which, when integrated over $v_{\|}$, produces

$$\operatorname{Re}\sigma=-\frac{1}{|k_{\|}|}\frac{\pi\omega\varepsilon_0\omega_p^2}{k^2}\sum_{\mathrm{n}=-\infty}^{\infty}\int 2\pi v_\perp dv_\perp \cdot\left[J_{\mathrm{n}}\left(\frac{k_\perp v_\perp}{\omega_c}\right)\right]^2\left[k_{\|}\frac{\partial f_0}{\partial v_{\|}}+\frac{\mathrm{n}\omega_c}{v_\perp}\frac{\partial f_0}{\partial v_\perp}\right]_{v_{\|}=(\omega-\mathrm{n}\omega_c)/k_{\|}} \tag{2.108}$$

To have a negative Re σ, one can choose $\partial f_0/\partial v_{\|}$ or $\partial f_0/\partial v_\perp$ to be positive. As we have seen in Section 2.2, $\partial f_0/\partial v_{\|}>0$ means a two stream in the direction parallel to the magnetic field; however the situation with $\partial f_0/\partial v_\perp>0$ is new. Hence we will treat this case in this subsection. We call this a type A distribution. Another possibility by which one can have a negative Re σ is for the combined quantities $k_{\|}\partial f_0/\partial v_{\|}+$ $+\mathrm{n}\,\omega_c/v_\perp\,\partial f_0/\partial v_\perp$ to be positive when evaluated at $v_{\|}=(\omega-\mathrm{n}\,\omega_c)/k_{\|}$. This happens even if f_0 is a decreasing function both in $v_\perp$ and $v_{\|}$, if we assume a range of ω such that $v_{\|}$ is negative. For example if we assume a symmetric velocity distribution in both $v_{\|}$ and $v_\perp$, we can write $\partial f_0/\partial v_{\|}=$ $=-v_{\|}/v_{T\|}^2 f_0$ and $\partial f_0/\partial v_\perp=-v_\perp/v_{T\perp}^2 f_0$ where $v_{T\|}$ and $v_{T\perp}$ are the root mean square velocities in the parallel and perpendicular directions. Then

$$-(k_{\|}\partial f_0/\partial v_{\|}+\mathrm{n}\,\omega_c/v_\perp\,\partial f_0/\partial v_\perp)\big|_{v_{\|}=(\omega-\mathrm{n}\omega_c)/k_{\|}}$$
$$=[\omega-\mathrm{n}\,\omega_c(1-v_{T\|}^2/v_{T\perp}^2)]f_0/v_{T\|}^2\,.$$

Therefore if $\omega<\mathrm{n}\,\omega_c(1-v_{T\|}^2/v_{T\perp}^2)$, we have a negative real conductivity. This requires, of course, $v_{T\|}/v_{T\perp}<1$, or an anisotropic temperature with $T_\perp>T_{\|}$. We will treat this case also in this subsection. We call this a type B distribution.

Let us first discuss electrostatic instabilities caused by $\partial f_0/\partial v_\perp>0$, the type A distribution. Because $f_0\to 0$ as $v_\perp\to\infty$, $\partial f_0/\partial v_\perp$ cannot be positive for the entire range of $v_\perp$. Hence this condition implies an existence of a hump in f_0 at $v_\perp\neq 0$, as shown in Fig. 14(A). The necessary condition for the instability is obtained from Eq. (2.28) by putting $k_{\|}=0$ and requiring the negative energy condition, $\partial\varepsilon/\partial\omega<0$, as

$$D_{\mathrm{n}}(k)\equiv\int_0^\infty\left[J_{\mathrm{n}}\left(\frac{k_\perp v_\perp}{\omega_c}\right)\right]^2\frac{1}{v_\perp}\frac{\partial f_0}{\partial v_\perp}2\pi v_\perp dv_\perp>0\,. \tag{2.109}$$

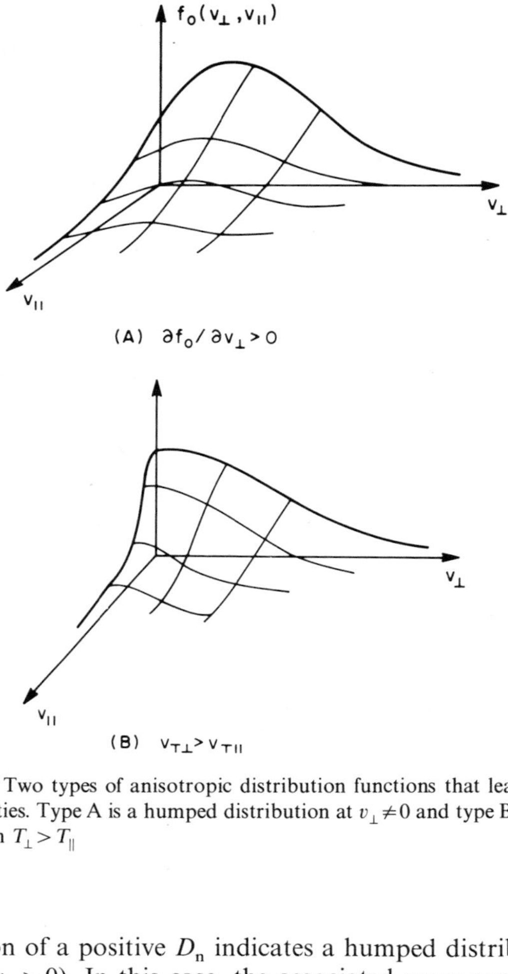

Fig. 14A and B. Two types of anisotropic distribution functions that lead to electrostatic plasma instabilities. Type A is a humped distribution at $v_\perp \neq 0$ and type B is an anisotropic distribution with $T_\perp > T_\parallel$

The condition of a positive D_n indicates a humped distribution in $v_\perp$ at $v_\perp \neq 0$ $(\partial f_0/\partial v_\perp > 0)$. In this case, the associated wave carries a negative energy when $k_\parallel = 0$; when $k_\parallel \neq 0$ the wave changes to the negative dissipative type because of the wave particle resonance. When $k_\parallel = 0$, instability occurs by the coupling between negative and positive energy branches of the dispersion relation of the same species, as well as by the coupling between a negative energy wave of one species and a positive energy wave of other species.

As an example let us introduce the case studied by Dory *et al.* (1965) (Dory-Guest-Harris instability). They have considered a flute mode $(k_\parallel = 0)$ excited by only one species. The dispersion relation can be obtained using the conductivity shown in Eq. (2.28) by taking $k_\parallel = 0$.

In particular, if we consider the contribution of only one species,* it reduces to

$$\frac{k^2}{\omega_p^2} = -\sum_{n=-\infty}^{\infty} \frac{n\,\omega_c}{\omega - n\,\omega_c} D_n(k), \tag{2.110}$$

D_n is defined in Eq. (2.109). If the $(n+1)^{st}$ harmonic has negative energy, $D_{n+1} > 0$, and the instability results by coupling into the positive energy branch of the nth harmonic; the corresponding frequencies lie in the range $n < \omega/\omega_c < n+1$. For $-1 < \omega/\omega_c < 1$, the conditions of the instability require $D_0 < 0$ and $D_1 > 0$. The real part of the unstable frequency is zero for this case. Dory *et al.* (1965) analysed this instability for the case where the perpendicular distribution function is given by

$$f_{0\perp}^{j}(v_\perp) \equiv \int_{-\infty}^{\infty} dv_{\|} f_0(v_\perp, v_{\|})$$

$$= \frac{1}{\pi\alpha^2 j!} \left(\frac{v_\perp}{\alpha}\right)^{2j} \exp\left(-v_\perp^2/\alpha^2\right) \tag{2.111}$$

with $j = 1, 2, 3 \ldots$. Such distributions are peaked at $\langle v_\perp \rangle = j^{1/2}\alpha$ and have a half width of approximately $\delta v_\perp \sim \alpha/(4j)^{1/2}$.

As the value of j increases, the distribution becomes more sharply peaked. They found no instability for $j = 0$, 1 and 2; the instability occurs for $j \geq 3$. The dependence of the threshold density n on the relative half width of the distribution is shown in Fig. 15. They found that when $j \geq 6$, the distribution sustains a growing wave whose real component of frequency is $\sim 1.2\,\omega_c$ with its density threshold given approximately by $\omega_p \sim 10\,\omega_c$. Instabilities generated by a humped distribution in $v_\perp$ have also been studied by Gruber *et al.* (1965) for a nonflute mode ($k_{\|} \neq 0$) and by Hall *et al.* (1965) for ion cyclotron waves by taking coupling to the electron mode into account.

Let us now look at the instability generated by a type B distribution (Fig. 14(B)) with a large perpendicular kinetic energy. This instability was first discussed by Harris (1961) and is often called the *Harris instability*. To study this instability let us take a distribution function which is Maxwellian in both the parallel and perpendicular directions but with $v_{T\perp} > v_{T\|}$;

$$f_0(v_{\|}, v_\perp) = \frac{1}{(2\pi)^{3/2}} \frac{1}{v_{T\|} v_{T\perp}^2} \exp\left[-\frac{v_{\|}^2}{2v_{T\|}^2} - \frac{v_\perp^2}{2v_{T\perp}^2}\right]. \tag{2.112}$$

* For electron mode, the ion contribution can be neglected if $\omega \gg \omega_{pi}$, while for ion mode, the electron contribution can be neglected if $\omega_{pe}^2/\omega_{ce}^2 \ll 1$.

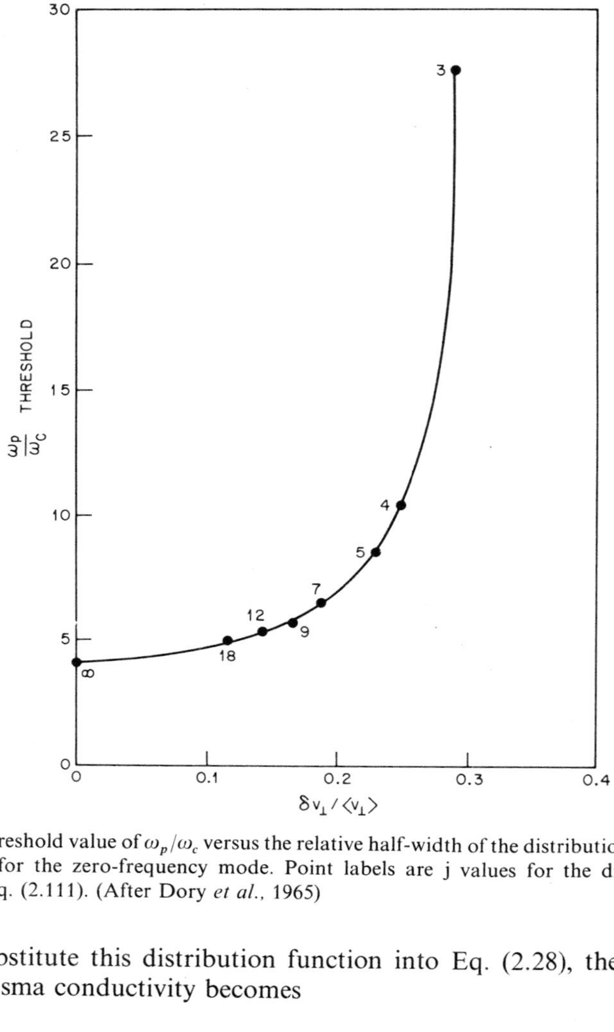

Fig. 15. Threshold value of ω_p/ω_c versus the relative half-width of the distribution function, $\delta v_\perp/\langle v_\perp \rangle$, for the zero-frequency mode. Point labels are j values for the distributions given by Eq. (2.111). (After Dory *et al.*, 1965)

If we substitute this distribution function into Eq. (2.28), the electrostatic plasma conductivity becomes

$$\sigma = -\frac{i\omega\varepsilon_0}{k^2 v_{T\|}^2}\,\omega_p^2 \sum_{n=-\infty}^{\infty} e^{-\lambda} I_n(\lambda) \left[1 + \frac{\omega - n\omega_c(1 - v_{T\|}^2/v_{T\perp}^2)}{\sqrt{2}\,k_\| v_{T\|}}\, Z\left(\frac{\omega - n\omega_c}{\sqrt{2}\,k_\| v_{T\|}}\right)\right] \tag{2.113}$$

where $\lambda = \left(\frac{k_\perp v_{T\perp}}{\omega_c}\right)^2$.

As it should, when the distribution is isotropic, $v_{T\|}=v_{T\perp}$, this expression reduces to the conductivity of the stable Bernstein wave, Eq. (2.30).

Unlike the previous case (type A), the conductivity shown in Eq. (2.113) does not admit a negative energy wave at $k_\|=0$. Hence the instability is of the negative dissipative type and occurs only for an obliquely propagating wave where $k_\| \neq 0$ as well as $k_\perp \neq 0$. The instability occurs again by coupling among different cyclotron harmonics of one species as well as coupling between a negative dissipative branch of one species and a positive energy wave of other species. Let us first study in what part of frequency range the conductivity (2.113) becomes negatively dissipative. Because the plasma dispersion function Z always has a positive imaginary part for a real argument, for $(\mathrm{n}-1)\,\omega_c < \omega < \mathrm{n}\,\omega_c$, one obvious condition is

$$\omega < \mathrm{n}\,\omega_c (1 - v_{T\|}^2 / v_{T\perp}^2) .$$

On the other hand when ω approaches $(\mathrm{n}-1)\,\omega_c$ from above, the cyclotron damping of the $(\mathrm{n}-1)^{\mathrm{st}}$ Bernstein wave tends to cancel the negative imaginary part of the nth harmonic. This can be shown to occur when $\omega = (\mathrm{n}-1/2)\,\omega_c$ (Soper and Harris, 1965). Thus the frequency range over which the conductivity becomes negatively dissipative is given by

$$\left(\mathrm{n}-\frac{1}{2}\right)\omega_c < \omega < \mathrm{n}\,\omega_c\left(1-\frac{v_{T\|}^2}{v_{T\perp}^2}\right). \tag{2.114}$$

From this expression we can immediately see that for an instability to occur,

$$\frac{v_{T\|}^2}{v_{T\perp}^2} = \frac{T_\|}{T_\perp} < \frac{1}{2\mathrm{n}} \tag{2.115}$$

and the unstable frequency lies around $(\mathrm{n}-1/2)\omega_c$, $\mathrm{n}=1, 2, \ldots$. This presents an interesting contrast with the previous case of $\partial f_0/\partial v_\perp > 0$, where the instability is expected around $\omega \sim \mathrm{n}\,\omega_c$.

In the magnetosphere, ring current protons and electrons may have distributions with a hump at $v_\perp \neq 0$ and $T_\perp > T_\|$. Although those particles alone are relatively stable unless the anisotropy is very large, in the region where cold plasma is intermixed, the instability may be generated by coupling to a wave carried by the cold plasma. To study this possibility, let us look at the dispersion relation of a cold plasma in a magnetic field. If we put $v_{T\|}=v_{T\perp}=0$ in Eq. (2.113), we can obtain the electrostatic conductivity for cold plasma σ_c,

$$\sigma_c = i\omega\varepsilon_0 \sum_{i,e} \left(\frac{k_{\parallel}^2}{k^2} \frac{\omega_p^2}{\omega^2} + \frac{k_{\perp}^2}{k^2} \frac{\omega_p^2}{\omega^2 - \omega_c^2} \right). \tag{2.116}$$

The dispersion relation of the wave carried by the cold plasma only is obtained by $-i\omega\varepsilon_0 + \sigma_c = 0$. Fig. 16 shows this dispersion relation plotted in the ω vs. $k_{\parallel}/k_{\perp}$ plane for the case $\omega_{pe} > \omega_{ce}$. In this figure ω_{UH} and ω_{LH} are the upper hybrid and the lower hybrid frequencies given by

$$\omega_{UH}^2 = \omega_{pe}^2 + \omega_{ce}^2, \tag{2.117}$$

$$\omega_{LH}^2 = \omega_{ci}^2 + \frac{\omega_{pi}^2}{1 + \omega_{pe}^2/\omega_{ce}^2}. \tag{2.118}$$

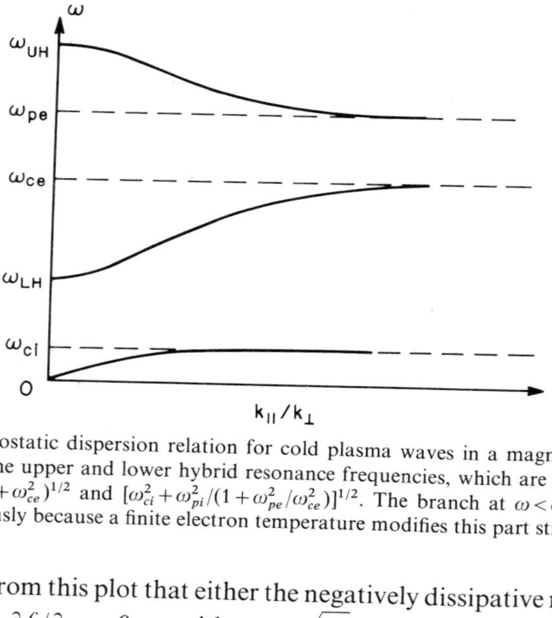

Fig. 16. Electrostatic dispersion relation for cold plasma waves in a magnetic field. ω_{UH} and ω_{LH} are the upper and lower hybrid resonance frequencies, which are approximately given by $(\omega_{pe}^2 + \omega_{ce}^2)^{1/2}$ and $[\omega_{ci}^2 + \omega_{pi}^2/(1 + \omega_{pe}^2/\omega_{ce}^2)]^{1/2}$. The branch at $\omega < \omega_{ci}$ should not be taken seriously because a finite electron temperature modifies this part strongly

We can see from this plot that either the negatively dissipative mode of hot plasma with $\partial f_0/\partial v_{\perp} > 0$, or with $v_{T\perp} > \sqrt{2}\, v_{T\parallel}$ at harmonics of cyclotron frequencies can couple to the cold plasma waves. The instability appears when the negatively dissipative mode couples with a positive energy wave. Recently abundant observations of electrostatic noise at frequencies of $n\omega_{ce}$ and $(n + 1/2)\omega_{ce}$ have been made by the OGO-5 satellite (Kennel *et al.*, 1970) in the equatorial region of the magnetosphere (Fig. 17). In particular the electric noise observed at $\omega_{ce}(n - 1/2)$, $n = 1, 2, \ldots$ is a

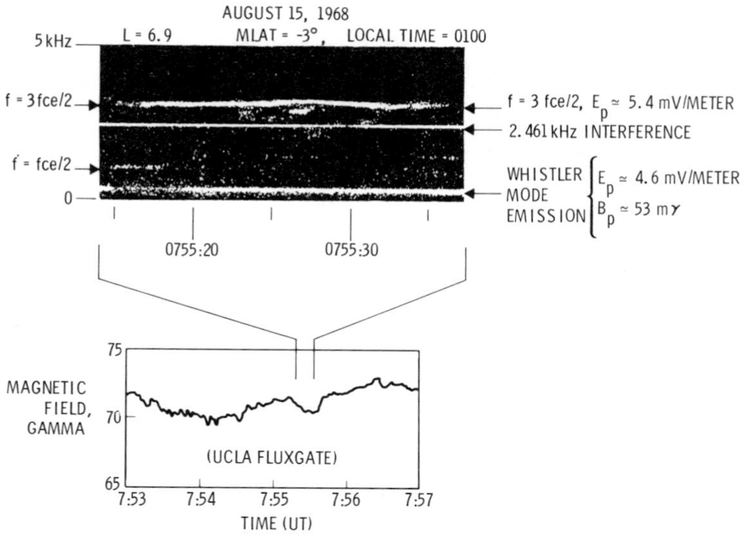

Fig. 17. Simultaneous occurrence of $f \approx 3f_c/2$ and whistler emissions. At the top is a 0 to 5 kHz spectrogram that contains an emission at $3f_c/2 \approx 3$ kHz, one near 0.5 kHz, and an interference time at 2.461 kHz from the spacercraft converter. The low-frequency emission proved to be electromagnetic. Simultaneous electromagnetic field data from the UCLA fluxgate magnetometer are also shown in the bottom inset. (After Kennel *et al.*, 1970)

strong indication of an electrostatic instability of the type discussed in this subsection.

In the absence of cold electrons, the type A distribution tends to be unstable at the zeroth harmonic of cyclotron frequencies, while the type B distribution is unstable for $n - 1/2 < \omega/\omega_{ce} < n$. Observed oscillations, say, at $3/2\ \omega_{ce}$ thus tend to indicate the presence of the Harris instability expected from a type B distribution. However the observed noise also appears frequently below $3/2\ \omega_{ce}$, where the type B distribution is stable. Consequently a more careful study of the instability should be made using a realistic distribution (a mixture of types A and B) and taking into account the cold electrons. Young *et al.* (1973) have made a very careful study to find a condition of instability which would explain the observed noise emission. They noted first that the observed electron distribution function at 6.6 R_E as measured by ATS-5, in fact has a mixed distribution of types A and B as shown in Fig. 18. Using this as an example of a realistic distribution function, they have concluded that a flute mode ($k_{\parallel} = 0$) is practically impossible to be the case. For $k_{\parallel} \neq 0$, they have shown that the main portion of the wave energy must come from a type A distribution rather than type B and the density ratio of cold to the

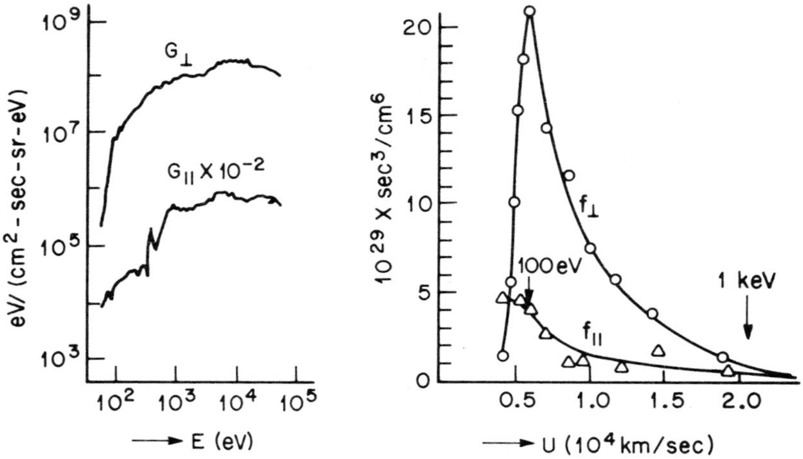

Fig. 18. Differential energy flux G from ATS-5 (De Forest and McIlwain, 1971) and the corresponding distribution function f of electrons. Subscripts $\perp$ and $\|$ represent the components measured perpendicular and parallel to the background magnetic field, respectively. The variable f is proportional to G divided by the square of the corresponding energy component. (After Young *et al*, 1973)

anisotropic hot component must be greater than 10^{-2} and that the temperature ratio of the two components for cases of high particle density must be no less than 0.1.

This is a good example of the use of the plasma instability theory together with the observed noise emissions to enable us to infer the actual particle velocity distribution function. If we consider a similar situation with respect to protons, the large ratio of ω_p/ω_c allows the instability conditions to be very easily satisfied even for less anisotropic distributions. This indicates that protons in the magnetosphere should have relatively isotropic distributions. When microinstabilities occur, the distribution function moves quickly to an isotropic form. To study the detailed process of the change of the distribution function, the technique of a computer simulation is often used. This is a technique in which the computer follows the dynamics of plasma particles exactly in the self generated field. Computer simulation of an electrostatic instability caused by anisotropic velocity distribution was tried first by Okuda and Hasegawa (1969). Fig. 19 shows the result obtained by Byers and Grewal (1970) for the humped distribution instability and Fig. 20 shows that of Gitomer *et al.* (1972) for the Harris instability. An interesting development of the distribution function can be seen in these figures, which indicate that a strong velocity space diffusion can result from these in-

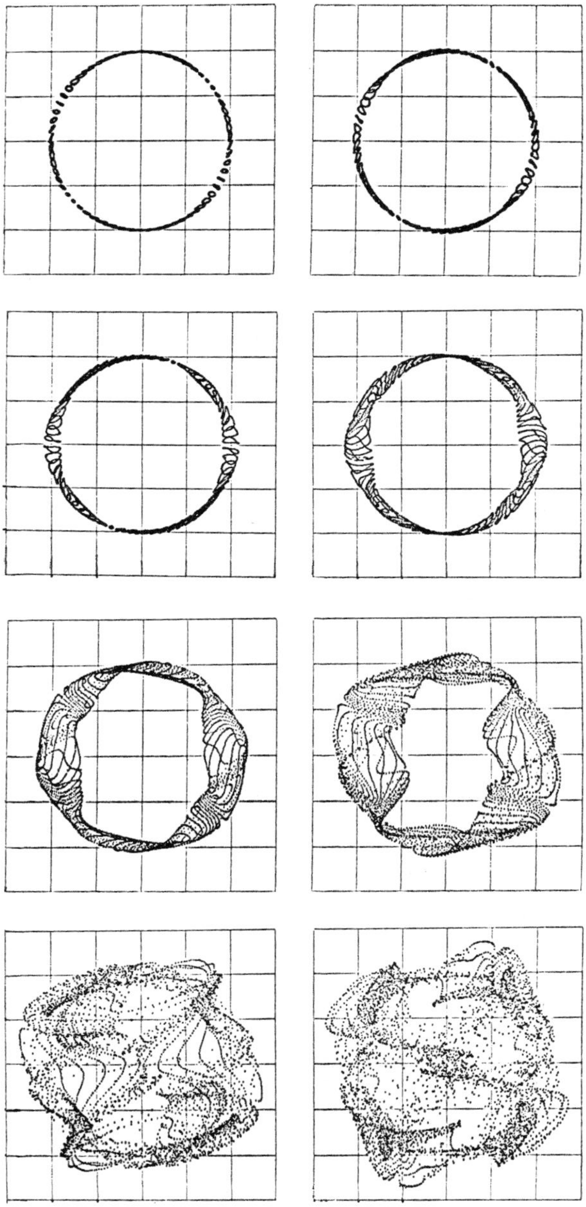

Fig. 19. Velocity-time development. The particles' positions in v_y, v_x space are shown for the following times (in the unit of $2\pi/\omega_c$) 1.0, 1.25, 1.5, 1.75, 2.0, 2.25, 2.5, 2.75. (After Byers and Grewel, 1970)

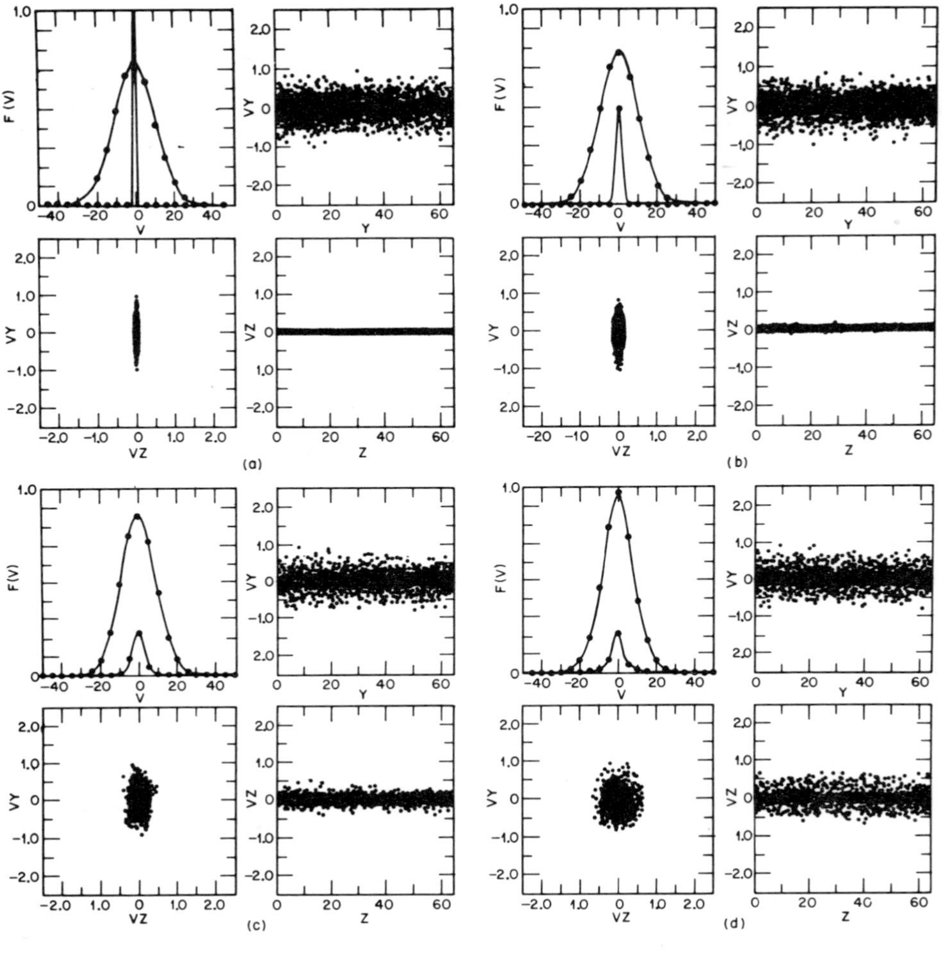

Fig. 20a–d. Plasma evolution for time 0.16 τ_c (a), 2.96 τ_c (b), 4.36 τ_c (c), 14.36 τ_c (d), ($\tau_c = 2\pi/\omega_c$). Displayed are v_y and v_z velocity histograms (upper left), $v_y - v_z$ velocity space (lower left), $z - v_z$ phase space (lower right, and $y - v_y$ phase space (upper right). Histogram horizontal axis units are 0.1 $(T_\perp(0)/\mathrm{m})^{1/2}$. Histogram vertical axis is the scaled number of particles (5000 = 1) for v_y and 0.05 times the scaled number of particles for v_z. In the velocity and phase-space plots, position is scaled by λ_D while velocity is scaled by $\lambda_D \omega_c$. Results quoted here are for the following set of parameters: $N = 96800$ particles, $\Delta z \times \Delta y = \lambda_D \times 2\lambda_D$, $\Delta t = 0.04\,\tau_c$, $z \times y$ mesh $= 64 \times 32$, $\omega_p/\omega_c = 1.0$, $T_\perp(0) = 10$ eV, and the number of simulation particles per Debye length squared ≈ 24. (After Gitomer *et al.*, 1972)

stabilities. When the distribution function is slightly unstable, there is a way to obtain analytically the diffusion process of the distribution function in velocity space. This will be discussed in Subsection 2.3d.

2.3b Electromagnetic Instabilities

Anisotropic velocity distributions can also cause electromagnetic perturbations to grow. The instability in this case can result either from a larger perpendicular temperature or from a larger parallel temperature. In the former case ($T_\perp > T_\parallel$), in contrast to the electrostatic case the instability occurs most easily for a wave propagating in the direction parallel to the magnetic field (Weibel, 1959). We discuss this instability in this subsection; we leave the other case to Section 2.4. We recall that when the direction of wave propagation is exactly parallel to the magnetic field, the electrostatic mode and the electromagnetic mode completely separate. The electromagnetic mode is called an electron or ion cyclotron wave depending on the direction of polarization.

When the parallel kinetic energy (temperature) approaches zero, the wave becomes the negative energy type and the instability occurs by a coupling with the particle cyclotron resonance. Let us first look at this ideal case for an example of the ion cyclotron wave. When the wave vector $\boldsymbol{k}$ is parallel to $\boldsymbol{B}_0$, we can use the dispersion relation derived in the previous section, Eq. (2.50). If we take a limit of zero temperature in the parallel direction, by substituting $\delta(v_\parallel)$ for $f_{0\parallel}(v)$ in this expression,

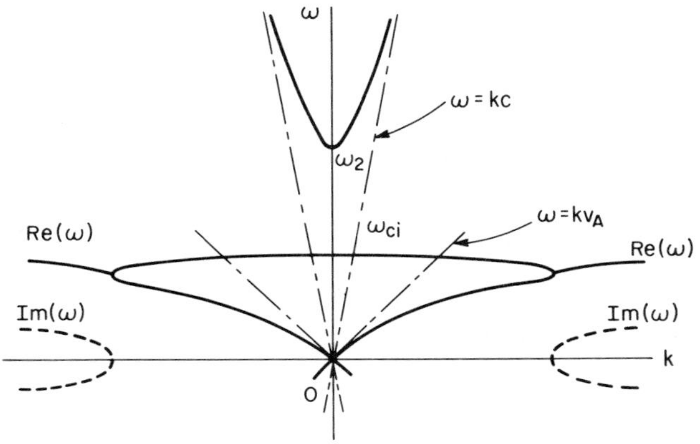

Fig. 21. Dispersion diagram of the ion-cyclotron wave in a plasma with anisotropic temperature ($T_\perp \gg T_\parallel$)

we obtain the dispersion relation,

$$k^2\left[1+\frac{\omega_{pe}^2\langle v_{\perp e}^2\rangle/2}{(\omega+\omega_{ce})^2 c^2}+\frac{\omega_{pi}^2\langle v_{\perp i}^2\rangle/2}{(\omega-\omega_{ci})^2 c^2}\right]=\frac{\omega^2}{c^2}\left[1-\frac{\omega_{pe}^2}{\omega(\omega+\omega_{ce})}-\frac{\omega_{pi}^2}{\omega(\omega-\omega_{ci})}\right]. \tag{2.119}$$

If we solve for the condition of a complex ω solution in this expression, we can obtain the range of the wave number k in which the instability occurs:

$$k>k_m=\left[\frac{\omega_{pi}\omega_{ci}}{c(\langle v_{\perp i}^2\rangle)^{1/2}}\right]^{1/2}.$$

For $k \gg k_m$ the growth rate Im ω and the corresponding real part of the frequency become

$$\operatorname{Im}\omega\approx\frac{\omega_{pi}}{c}\left(\frac{\langle v_{\perp i}^2\rangle}{2}\right)^{1/2} \tag{2.120}$$

and

$$\operatorname{Re}\omega\approx\omega_{ci}.$$

The dispersion relation (2.119) is plotted in Fig. 21.

In the above example we have taken an ideal case with $T_\perp/T_\parallel\to\infty$ to demonstrate the instability. If we introduce a finite temperature in $T_\parallel$, an immediate complication arises in the dispersion relation because of the singularity $(\omega-\omega_c-k_\parallel v_\parallel=0)$ in the velocity integral which produces cyclotron damping. The instability becomes the negatively dissipative type. If the growth rate is small, we can use expression (1.69) to obtain the growth rate:

$$\omega_i=-\frac{\operatorname{Im}D}{\partial\operatorname{Re}D/\partial\omega}. \tag{1.69}$$

For the ion cyclotron wave, D is obtained from Eq. (2.50),

$$D_i=k^2c^2-\omega^2+\omega_{pe}^2\int\frac{(\omega-kv_\parallel)f_{0\parallel}^{(e)}(v_\parallel)-k(\langle v_{\perp e}^2\rangle/2)\,(\partial f_{0\parallel}^{(e)}/\partial v_\parallel)}{\omega-kv_\parallel+\omega_{ce}}\,dv_\parallel$$

$$+\omega_{pi}^2\int\frac{(\omega-kv_\parallel)f_{0\parallel}^{(i)}(v_\parallel)-k(\langle v_{\perp i}^2\rangle/2)\,(\partial f_{0\parallel}^{(i)}/\partial v_\parallel)}{\omega-kv_\parallel-\omega_{ci}}\,dv_\parallel. \tag{2.121}$$

For this choice of D, $\partial \mathrm{Re} D/\partial\omega < 0$ corresponds to a positive energy wave.* Let us now calculate $\mathrm{Im}\, D_i$ from the integral around the pole $k v_{\|} = \omega - \omega_{ci}$. If we ignore contribution of electrons,

$$\mathrm{Im}\, D_i = \frac{-\pi\omega_{pi}^2}{|k|}\left[\omega_{ci} f_{0\|}^{(i)}(v_{\|}) - k(\langle v_{\perp i}^2\rangle/2)(\partial f_{0\|}^{(i)}/\partial v_{\|})\right]$$

at $k v_{\|} = \omega - \omega_{ci}$.

For a Maxwellian distribution, $\partial f_0/\partial v_{\|} = -v_{\|}/v_{T\|}^2 f_0$, while $\langle v_{\perp}^2\rangle/2 = v_{T\perp}^2$. Hence if we write $v_{T\perp}^2/v_{T\|}^2 = T_{\perp}/T_{\|}$, we have

$$\mathrm{Im}\, D_i = -\frac{\pi\omega_{pi}^2}{|k|}\left[\omega_{ci} - (\omega_{ci} - \omega)\frac{T_{\perp i}}{T_{\| i}}\right] f_0^{(i)}\left(\frac{\omega - \omega_{ci}}{k}\right). \quad (2.122)$$

Similarly for the electron cyclotron wave we have, from Eq. (2.49),

$$\mathrm{Im}\, D_e = -\frac{\pi\omega_{pe}^2}{|k|}\left[\omega_{ce} - (\omega_{ce} - \omega)\frac{T_{\perp e}}{T_{\| e}}\right] f_0^{(e)}\left(\frac{\omega - \omega_{ce}}{k}\right).^{**} \quad (2.123)$$

As we have seen the cyclotron wave becomes of the negative energy type at the limit of $T_{\perp}/T_{\|} \to \infty$. Consequently, the instability cannot be decided by the sign of $\mathrm{Im}\, D$ alone. However, if a background medium carries a positive energy wave in the frequency range where $\mathrm{Im}\, D > 0$ (negative dissipation), we will have an instability. Again this situation is provided by the existence of a background cold plasma. When the cold plasma density exceeds the hot plasma density *significantly* the real part of D is essentially decided by the cold plasma dielectric constant:

$$D_{\substack{i\\e}} \sim k^2 c^2 - \omega^2 + \frac{\omega\omega_{pe}^2}{\omega \pm \omega_{ce}} + \frac{\omega\omega_{pi}^2}{\omega \mp \omega_{ci}} \quad (2.124)$$

and

$$\partial \mathrm{Re} D/\partial\omega < 0 .$$

Then the instability condition is given by $\mathrm{Im}\, D > 0$, or from Eq. (2.122) and (2.123),

$$\frac{T_{\perp}}{T_{\|}} > \frac{\omega_c}{\omega_c - \omega}. \quad (2.125)$$

---

* This can be seen from the portion of the dispersion relation that shows the electromagnetic wave in vacuum; $k^2 c^2 - \omega^2$ and by taking the derivative with respect to ω.

** Contribution of the poles at $(\omega + \omega_{cj})/k$, $j = i, e$, was ignored.

This expression shows that even for a small anisotropy in temperature, an instability can result for $\omega \to 0$.

That the electrons in the radiation belt may be subject to the cyclotron-wave instability because of their anisotropic pitch-angle distribution was first pointed out by Brice (1963). The idea was developed by Kennel and Petschek (1966), who calculated the limit of stably trapped particle fluxes with respect to the cyclotron instabilities. This will be discussed later in Subsection 2.3d.

If we combine Eq. (2.125) with the resonant condition $\omega - \omega_c = k v_{||}$ and eliminate k using the dispersion relation of cyclotron waves in the cold plasma ($D=0$ of Eq. (2.124)), we can calculate the energy E_R of the resonant "hot" particles that contribute to the instability from $E_R = m v_{||}^2/2$ as:

$$E_{Re} = E_B \frac{(T_{||e}/T_{\perp e})^3}{1 - T_{||e}/T_{\perp e}}, \tag{2.126}$$

$$E_{Ri} = E_B \frac{(T_{||i}/T_{\perp i})^3}{(1 - T_{||i}/T_{\perp i})^2} \tag{2.127}$$

where $[E_B = B_0^2/(2\mu_0 n_{0c})]$ is the magnetic field energy per particle of *cold* plasma that carries the wave.

Note the difference in the power in the denominator which results in a much larger resonant energy for protons than for electrons.

E_B, in units of eV, can be expressed in terms of the magnetic-flux density B_0 in γ and the cold electron density n_{0c} in cm^{-3},

$$E_B(\mathrm{eV}) = \frac{B_0^2(\gamma)}{0.4\, n_{0c}(\mathrm{cm}^{-3})}. \tag{2.128}$$

Thus, say, for $B_0 \sim 200\,\gamma$, $n_{0c} = 1\ \mathrm{cm}^{-3}$, E_B is 100 keV. Therefore, for this example, if the anisotropy, $T_{||}/T_{\perp}$, is 0.8, 250-keV electrons and 1250-keV protons resonate with the wave, lose energy, and precipitate through a pitch-angle diffusion.

However, one can see from expression (2.128) that the energy of resonant hot particles can be significantly reduced if n_{0c} is increased, say, by a factor of 10^2. This occurs inside the plasmapause, leading to the expectation that an anomalous loss of particles inside the plasmapause may occur (Cornwall *et al.*, 1970).

Fukunishi (1969) explained sweepers (geomagnetic pulsations in the pc 1 range that are characterized by broad band noise with gradually increasing frequency) observed in the evening sector in terms of the proton cyclotron instability: he postulated that injected protons from the geomagnetic tail drift toward the west due to curvature and gradient of the earth's magnetic field with a drift speed which is proportional to the proton energy. Hence higher energy protons arrive at the observation point earlier and emit lower frequency noise which, upon arrival of lower energy protons, gradually changes to higher frequency.

Thorne (1968) considered an obliquely propagating whistler wave in which the Landau pole ($\omega = k_{\parallel} v_{\parallel}$ resonance) contribution becomes important. As was discussed for the electrostatic case $\partial f_0/\partial v_{\parallel}$ must be positive at the Landau resonant velocity ($v_{\parallel} = \omega/k_{\parallel}$) in order to have $\mathrm{Re}\,\sigma < 0$, rather than the anisotropy needed for the cyclotron-resonant velocity. Thorne has shown that an unducted whistler propagating obliquely to the magnetic field may be amplified by the secondary peak ($\partial f_0/\partial v_{\parallel} > 0$) in the energy distribution around 10 keV.

We conclude this section by stating that the cyclotron-wave instability serves as an important loss mechanism for radiation-belt particles. However, because a large anisotropy is required, this mechanism is less effective for ring-current particles except possibly inside the plasmapause.

2.3c Effect of Nonuniformity on Anisotropic Distributions

When the plasma density n_0 is not uniform in space, a diamagnetic current $\boldsymbol{J}_d = -\nabla(n_0 T) \times \boldsymbol{B}/B^2$ is generated in the plasma and a perturbation in the density on this current will have a Doppler-shifted frequency $\omega^* = k_{\perp} v_d$, where $e n_0 v_d = J_d$. Consequently, in a nonuniform plasma, a new mode whose dispersion relation is given by $\omega = \omega^*$ is generated. This new mode is called "drift wave". In addition, if the plasma admits a wave whose phase velocity is smaller than v_d, such a wave will see a positive gradient part of the particle distribution function ($\partial f_0/\partial v > 0$) that participates in the diamagnetic current, hence the dielectric constant for this wave becomes negatively dissipative. This effect causes a new instability called "drift wave instability".

We will treat this instability in Section 4.1. Here, instead, we consider how the existence of a wave due to the nonuniformity contributes to the instability generated by anisotropic distributions. To treat waves in a nonuniform plasma, we must solve the Vlasov equation taking into account the effect of nonuniformity. This often leads to unnecessary complications. However, when the wave frequency ω of interest is much smaller than the cyclotron frequency ω_c and the wavelength of interest k^{-1} is much longer than the Larmor radius v_T/ω_c, we can average the Vlasov equation over space and time and take only low frequency and long wavelength contributions. The continuity equation in phase space derived this way is called the *Drift Kinetic Equation* and has the following general form

$$\frac{\partial f_D}{\partial t}+\nabla\cdot(\boldsymbol{v}_D f_D)+\frac{\partial}{\partial v_{||}}\left(\frac{F_{||}}{m}f_D\right)=0, \tag{2.129}$$

where f_D is the "averaged" distribution function which depends on $v_{||}$, $\mu\,(=mv_\perp^2/2B_0)$, $\boldsymbol{x}$ and t i.e.,

$$f_D=f_D(v_{||},\mu,\boldsymbol{x},t). \tag{2.130}$$

In this expression, $\boldsymbol{v}_D$ is the particle drift velocity which is given by

$$\boldsymbol{v}_D=\boldsymbol{v}_{D\perp}+\boldsymbol{v}_{D||}, \tag{2.131}$$

$$\boldsymbol{v}_{D||}=v_{||}\boldsymbol{B}/B, \tag{2.132}$$

$$\boldsymbol{v}_{D\perp}=\boldsymbol{v}_E+\frac{\boldsymbol{F}\times\boldsymbol{B}}{qB^2}. \tag{2.133}$$

$\boldsymbol{v}_E$ is the drift velocity due to an electric field given by

$$\boldsymbol{v}_E=\frac{(\boldsymbol{E}+\boldsymbol{v}_{||}\times\boldsymbol{B})\times\boldsymbol{B}}{B^2} \tag{2.134}$$

and $\boldsymbol{F}_\perp$ is the non-electric force acting on the particle, given by

$$\boldsymbol{F}_\perp=-\mu\nabla_\perp B-\frac{mv_{||}^2}{R}\hat{\boldsymbol{R}}-m\frac{d\boldsymbol{v}_E}{dt}+m\boldsymbol{g}. \tag{2.135}$$

In Eq. (2.134), $\boldsymbol{v}_{||}\times\boldsymbol{B}$ may look like zero but must be included when we discuss electromagnetic waves where $\boldsymbol{B}$ represents a wave magnetic field perpendicular to the ambient field $\boldsymbol{B}_0$. In Eq. (2.135) the first term produces a drift due to the gradient of the magnetic flux density, the second term drift due to curvature of the magnetic field ($\hat{\boldsymbol{R}}$ is the unit vector directed toward center of the curvature), the third term produces polarization drift and the last term gravitational drift. $\boldsymbol{F}_{||}$ is the force acting in the parallel direction given by

$$\boldsymbol{F}_{||}=-\mu\nabla_{||}B+q\boldsymbol{E}_{||}. \tag{2.136}$$

Let us now use this drift kinetic equation to calculate the equivalent conductivity for a species with charge q and mass m when the plasma density is nonuniform. Let us consider an electrostatic perturbation ($\boldsymbol{E} = -\nabla\varphi$) which propagates in a plane perpendicular to the direction of nonuniformity. We assume that the magnetic field is straight so that the curvature drift and the mirror force (the first term in Eq. (2.136)) are absent. Furthermore ∇B drift is smaller than $\boldsymbol{E}\times\boldsymbol{B}$ drift by a factor of β $(=2\mu_0 n_0 T/B_0^2)$, thus for a low β plasma the only important drifts are the electric field drift and the polarization drift;

$$\boldsymbol{v}_{D\perp} = \frac{1}{B_0^2}\left[\boldsymbol{E} - \frac{m}{qB_0^2}\left(\frac{d\boldsymbol{E}}{dt}\times\boldsymbol{B}_0\right)\right]\times\boldsymbol{B}_0 = \frac{\boldsymbol{E}\times\boldsymbol{B}_0}{B_0^2} + \frac{m}{qB_0^2}\frac{d\boldsymbol{E}}{dt}. \tag{2.137}$$

If we use $\boldsymbol{E}_1 = -\nabla\varphi_1 = -i(k_\perp\hat{\boldsymbol{y}} + k_\parallel\hat{\boldsymbol{z}})\varphi_1$, where $\hat{\boldsymbol{z}}$ is the direction of the ambient magnetic field and $\hat{\boldsymbol{y}}$ is that perpendicular to the direction of nonuniformity, and note that $d/dt = -i(\omega - k_\parallel v_\parallel)$, we can obtain the solution of the linearized distribution function f_1 of the drift kinetic equation as

$$f_1 = \left[\frac{\dfrac{q}{m}\, ik_\parallel \dfrac{\partial f_0}{\partial v_\parallel} + \dfrac{ik_\perp}{B_0}\dfrac{\partial f_0}{\partial x}}{i(k_\parallel v_\parallel - \omega)} - \frac{m}{qB_0^2}k_\perp^2 f_0\right]\varphi_1 . \tag{2.138}$$

When f_0 is a Maxwellian in $v_\parallel$ the equivalent conductivity can be expressed as

$$\sigma = \frac{-i\omega\varepsilon_0\omega_p^2}{k^2 v_T^2}\left[1 + \frac{\omega - \omega^*}{\sqrt{2}\,k_\parallel v_T} Z\left(\frac{\omega}{\sqrt{2}\,k_\parallel v_T}\right)\right] - i\omega\varepsilon_0\frac{\omega_p^2}{\omega_c^2}\frac{k_\perp^2}{k^2} \tag{2.139 a}$$

where ω^* is the drift wave frequency defined by

$$\omega^* = \frac{k_\perp v_T^2}{\omega_c}\frac{\partial f_0}{\partial x}\frac{1}{f_0} = -\frac{k_\perp \kappa v_T^2}{\omega_c}. \tag{2.140}$$

$$\kappa = -\partial(\ln f_0)/\partial x$$

(We choose a sign convention such that $\omega^* > 0$ for electrons for a positive κ.) If we compare the above conductivity with that for a uniform plasma obtained in Eq. (2.113) (by taking only $n=0$ term), we can see that they are the same except that ω which is multiplied by Z functions in the uniform plasma is replaced by $\omega - \omega^*$. That is, the frequency is Doppler shifted by an amount $k_\perp v_d$, where v_d is the effective drift speed of particles due to the diamagnetic current. Note however that v_d is *not* the real drift speed of particles, but is an "effective drift" in that $q n_0 v_d$ gives the diamagnetic current J_d due to the plasma nonuniformity. The reader is recommended to consult an elementary plasma physics text to clarify this point.

As was mentioned previously in this subsection, the real part of the conductivity has a negative value for $\omega < \omega^*$, which is responsible for causing the drift wave instability.

As stated before, we will not treat the drift wave instability here but will discuss the effect of a newly introduced mode on the instability due to the velocity space anisotropy. To study this problem, let us take a flute mode, $k_{\|}=0$, and look at the electron mode first. From Eq. (2.139a), when $k_{\|}=0$, the electron conductivity becomes,

$$\sigma_e = -i\omega\varepsilon_0 \left(\frac{\omega_{pe}^2}{\omega_{ce}^2} - \frac{\omega_{pe}^2}{\omega\omega_{ce}} \frac{\kappa}{k_\perp} \right). \tag{2.141}$$

If we ignore temporarily the ion dynamics, the dispersion relation, $-i\omega\varepsilon_0 + \sigma_e = 0$, gives

$$\omega = \frac{\omega_{pe}^2}{\omega_{ce}} \frac{\kappa}{k_\perp} \frac{1}{1+\omega_{pe}^2/\omega_{ce}^2}. \tag{2.142}$$

Therefore, we can see that in a nonuniform plasma, electrons carry a new mode at $\omega \sim \omega_{pe}\kappa/k_\perp \ll \omega_{pe}$. This new mode (not $\omega = \omega^*$ in this case)* can couple with an ion Bernstein wave with an anisotropic velocity distribution and can cause a new instability. Note that if $\kappa = 0$, this mode disappears and it takes a much larger anisotropy of ions to produce an instability by coupling only among ion modes (as shown in the case of Dory-Guest-Harris instability).

The instability generated by this newly introduced mode was found by Post and Rosenbluth (1966) and is often called the "drift-cone instability", because they used a loss-cone velocity distribution for ions.

Let us derive this drift cone instability. The instability is generated by a loss cone distribution of ions which is characterized by $f_0(v_\perp = 0) = 0$, hence it can be applied to ring current protons in the magnetosphere. The instability occurs for a short wavelength perturbation such that $k_\perp v_{Ti}/\omega_{ci} \gg 1$. Note that we assumed, however, $k_\perp v_{Te}/\omega_{ce} \ll 1$ for electrons in deriving Eq. (2.141).

* If ions are cold (i.e., $k_\perp v_{Ti} \ll \omega_{ci}$), the second term in Eq. (2.141) cancels with the corresponding ion term and we must retain the next order term which is proportional to v_{Te}^2. This term produces the drift wave.

For such a short wavelength region, the contribution of the term arising from the density gradient (2nd term in Eq. (2.141) for electrons) becomes negligible for ions because it is multiplied by a factor of $\omega_{ci}/k_\perp v_{Ti}$. Then the ion conductivity can be obtained from Eq. (2.28) by putting $k_{||}=0$

$$\sigma_i = -\frac{i\omega\varepsilon_0\omega_{pi}^2}{k_\perp^2}\sum_{n=-\infty}^{\infty}\int 2\pi\, dv_\perp^2 J_n^2\left(\frac{k_\perp v_\perp}{\omega_{ci}}\right)\frac{n\omega_{ci}\partial f_0/\partial v_\perp^2}{\omega - n\omega_{ci}}. \tag{2.143a}$$

If we write $n\omega_{ci}/(\omega - n\omega_{ci}) = -1 + \omega/(\omega - n\omega_{ci})$ and note that $\sum_n J_n^2 = 1$, as well as $f_0(v_\perp = 0) = 0$, we have, by integration by parts,

$$\begin{aligned}\sigma_i &= -\frac{i\omega\varepsilon_0\omega_{pi}^2}{k_\perp^2}\sum_{n=-\infty}^{\infty}\int 2\pi\, dv_\perp^2 J_n^2 \frac{\omega}{\omega - n\omega_{ci}}\frac{\partial f_0}{\partial v_\perp^2}\\ &= \frac{i\omega\varepsilon_0\omega_{pi}^2}{k_\perp^2}\sum_{n=-\infty}^{\infty}\int 2\pi\, dv_\perp^2 \frac{\omega}{\omega - n\omega_{ci}}\frac{\partial J_n^2}{\partial v_\perp^2} f_0 .\end{aligned}$$

Furthermore, if we assume short wavelengths, $J_n^2 \sim \omega_{ci}/(k_\perp v_\perp \pi)$, and we have

$$\sigma_i = \frac{-i\omega\varepsilon_0\omega_{pi}^2\omega_{ci}}{\pi k_\perp^3 v_{Ti}^3}\sum_{n=-\infty}^{\infty}\frac{\omega}{\omega - n\omega_{ci}} \tag{2.143b}$$

where we defined

$$v_{Ti}^{-3} = \int \frac{f_0}{v_\perp^3} 2\pi v_\perp\, dv_\perp .$$

If we combine the ion conductivity with the electron conductivity we obtain the following dispersion relation:

$$1 + \frac{\omega_{pe}^2}{\omega_{ce}^2} = \frac{\omega_{pe}^2\kappa}{\omega\omega_{ce}k_\perp} - \frac{\omega_{pi}^2\omega_{ci}}{k_\perp^3 v_{Ti}^3\pi}\sum_{n=-\infty}^{\infty}\frac{\omega}{\omega - n\omega_{ci}}. \tag{2.144}$$

It is not difficult to prove that the above dispersion relation contains unstable roots near harmonics of the ion cyclotron frequency. The unstable roots appear because of the coupling between the negative energy branch of the ion Bernstein mode and the electron drift mode and they disappear when $\kappa \to 0$. This instability may have an important application to the loss of ring current protons in outside the plasmapause. Inside the plasmapause the cold ions fill up the loss cone and thus stabilize this instability.

2.3d Quasi-Linear Diffusion in Velocity Space

When a microinstability develops, fluctuating fields generated by the instability scatter plasma particles and cause diffusion of the velocity distribution function. This can be regarded as an elementary process of induced scattering of a particle with momentum $\boldsymbol{p}$ by a wave with momentum $\hbar\boldsymbol{k}$;

$$\boldsymbol{p}=\boldsymbol{p}'+\hbar\boldsymbol{k}.$$

If the number density of a wave quantum $N_{\boldsymbol{k}}$ (= wave energy density/$\hbar\omega$) is large enough so that spontaneous excitation is negligible, the change of the particle distribution function in momentum space can be written as

$$\begin{aligned}\frac{\partial f_{\boldsymbol{p}}}{\partial t}=\sum_{\boldsymbol{k}}N_{\boldsymbol{k}}\{&w_{\boldsymbol{p}+\Delta\boldsymbol{p}}(f_{\boldsymbol{p}+\Delta\boldsymbol{p}}-f_{\boldsymbol{p}})\\&-w_{\boldsymbol{p}}(f_{\boldsymbol{p}}-f_{\boldsymbol{p}-\Delta\boldsymbol{p}})\}\end{aligned}\tag{2.145}$$

where $w_{\boldsymbol{p}}$ shows the transition probability and $\Delta\boldsymbol{p}=\hbar\boldsymbol{k}$. Because the wave momentum in the non-optical frequency range is much smaller than the particle momentum* we can assume $\boldsymbol{p}\gg\Delta\boldsymbol{p}$ and expand both $f_{\boldsymbol{p}}$ and $w_{\boldsymbol{p}}$ in the powers of $\Delta\boldsymbol{p}$.

$$\begin{aligned}\frac{\partial f_{\boldsymbol{p}}}{\partial t}&=\sum_{\boldsymbol{k}}N_{\boldsymbol{k}}\left(\frac{\partial f_{\boldsymbol{p}}}{\partial p_i}\Delta p_i\frac{\partial w_{\boldsymbol{p}}}{\partial p_j}\Delta p_j+\Delta p_i\,\Delta p_j\frac{\partial^2 f_{\boldsymbol{p}}}{\partial p_i\partial p_j}w_{\boldsymbol{p}}\right)\\&\equiv\frac{\partial}{\partial p_i}\left(D_{ij}\frac{\partial f_{\boldsymbol{p}}}{\partial p_j}\right)\end{aligned}\tag{2.146}$$

where

$$D_{ij}=\sum_{\boldsymbol{k}}\Delta p_i\,\Delta p_j\,N_{\boldsymbol{k}}\,w_{\boldsymbol{p}}\,.$$

The above result shows that such a process in fact leads to diffusion of the momentum (velocity) distribution function.

Here we will study such a diffusion process for two typical examples: the electrostatic and electromagnetic instabilities excited by anisotropic

* $\dfrac{\hbar k}{p}\sim\dfrac{\hbar k}{m v_T}=\dfrac{v_T}{v_p}\dfrac{\hbar\omega}{T}$, where v_T is the particle thermal speed, v_p is the wave phase speed, T is the plasma temperature in energy unit and ω is the angular frequency of the wave. Note ω which corresponds to 1 eV energy is $\sim10^{15}\ \mathrm{sec}^{-1}$.

velocity distributions. In both cases, the existence of a magnetic field in the plasma produces additional quantization of wave energy, i.e. $\hbar\omega_c$, hence, the change of particle monumentum $\Delta \boldsymbol{p}$ becomes momentum dependent. This makes $\Delta \boldsymbol{p}$ and $\partial/\partial \boldsymbol{p}$ not commutable and consequently the diffusion equation has a somewhat different form than Eq. (2.146).

In the case of an electrostatic instability, the quasi-linear diffusion in velocity space can be written from the Vlasov equation as (Sec. 4.2).

$$\frac{\partial f_0}{\partial t} = \mathrm{Re}\left[\frac{q}{m}\sum_{\boldsymbol{k}}(i\boldsymbol{k}\varphi_{\boldsymbol{k}}^{(1)})^* \cdot \frac{\partial f_{\boldsymbol{k}}^{(1)}}{\partial \boldsymbol{v}}\right]. \tag{2.147}$$

Here q is the charge of the particle, $\varphi_{\boldsymbol{k}}^{(1)}$ is the amplitude of the electrostatic potential of the excited wave, Re shows the real part and $f_{\boldsymbol{k}}^{(1)}$ is the perturbed velocity distribution function associated with $\varphi_{\boldsymbol{k}}^{(1)}$ which is given from the linearized Vlasov equation as

$$f_{\boldsymbol{k}}^{(1)} = \frac{q}{m}\varphi_{\boldsymbol{k}}^{(1)}\sum_{\mathrm{m,n}} J_{\mathrm{n}}\left(\frac{k_\perp v_\perp}{\omega_c}\right) J_{\mathrm{m}}\left(\frac{k_\perp v_\perp}{\omega_c}\right) \frac{\dfrac{\mathrm{n}\omega_c}{v_\perp}\dfrac{\partial f_0}{\partial v_\perp} + k_{||}\dfrac{\partial f_0}{\partial v_{||}}}{k_{||}v_{||} + \mathrm{n}\omega_c - \omega}\, \mathrm{e}^{i(\mathrm{m}-\mathrm{n})\theta}. \tag{2.148}$$

In this expression, $v_\perp$ and $v_{||}$ are perpendicular and parallel velocities, $\theta(=\tan^{-1} v_y/v_x)$ is the phase angle in v_x and v_y (velocity coordinates perpendicular to the ambient magnetic field B_0), $\omega_c = qB_0/m$ and J_{n} is the Bessel function. If we substitute Eq. (2.148) into (2.147) and average over the phase angle θ by integrating $\int_0^{2\pi} d\theta$, we have;

$$\frac{\partial f_0(v_\perp, v_{||})}{\partial t} = 2\pi^2\left(\frac{q}{m}\right)^2 \sum_{\boldsymbol{k}} |\varphi_{\boldsymbol{k}}^{(1)}|^2 \sum_{\mathrm{n}=-\infty}^{\infty}$$

$$\left(\frac{\mathrm{n}\omega_c}{v_\perp}\frac{\partial}{\partial v_\perp} + k_{||}\frac{\partial}{\partial v_{||}}\right) J_{\mathrm{n}}^2\left(\frac{k_\perp v_\perp}{\omega_c}\right) \tag{2.149}$$

$$\frac{1}{|k_{||}|}\delta\left(v_{||} - \frac{\omega - \mathrm{n}\omega_c}{k_{||}}\right)\left(\frac{\mathrm{n}\omega_c}{v_\perp}\frac{\partial f_0}{\partial v_\perp} + k_{||}\frac{\partial f_0}{\partial v_{||}}\right).$$

In the above derivations we assumed a small growth rate so that

$$\mathrm{Im}\left[\frac{1}{k_{||}v_{||} + \mathrm{n}\omega_c - \omega}\right] = \frac{\pi}{|k_{||}|}\delta\left(v_{||} - \frac{\omega - \mathrm{n}\omega_c}{k_{||}}\right). \tag{2.150}$$

For the case of electromagnetic cyclotron waves (propagating parallel to the magnetic field) we use left and right hand polarized electric field vectors E_L and E_R. Following a similar argument, we can derive the quasilinear equation for this case

$$\begin{aligned}\frac{\partial f_0(v_\perp, v_\parallel)}{\partial t} = 2\pi^2 \left(\frac{q}{m}\right)^2 \sum_k \Bigg[\frac{|E_{L,k}|^2}{2}\left\{\left(1-\frac{k_\parallel v_\parallel}{\omega}\right)\right. \\ \cdot \left.\frac{1}{v_\perp}\frac{\partial}{\partial v_\perp}v_\perp + \frac{k_\parallel v_\perp}{\omega}\frac{\partial}{\partial v_\parallel}\right\}\frac{1}{|k_\parallel|}\delta\left(v_\parallel - \frac{\omega-\omega_c}{k_\parallel}\right) \\ + \frac{|E_{R,k}|^2}{2}\left\{\left(1-\frac{k_\parallel v_\parallel}{\omega}\right)\frac{1}{v_\perp}\frac{\partial}{\partial v_\perp}v_\perp + \frac{k_\parallel v_\perp}{\omega}\frac{\partial}{\partial v_\parallel}\right\} \\ \cdot \frac{1}{|k_\parallel|}\delta\left(v_\parallel - \frac{\omega+\omega_c}{k_\parallel}\right)\Bigg\}\Bigg]\left\{\left(1-\frac{k_\parallel v_\parallel}{\omega}\right)\frac{\partial f_0}{\partial v_\perp} + \frac{k_\parallel v_\perp}{\omega}\frac{\partial f_0}{\partial v_\parallel}\right\}.\end{aligned} \quad (2.151)$$

In these expressions ω_c includes a sign, so that for an electron cyclotron wave $\omega_c = -\omega_{ce}$ (<0).

Let us now see the consequence of these diffusion processes. Obviously $f_0(v_\perp, v_\parallel)$ changes slowly due to the fluctuating fields, but in what way? One simple answer to this question is to look for the contour in velocity space along which the diffusion takes place. In the case of the cyclotron wave, this can be found from Eq. (2.151). In view of the δ function, if we assume a narrow spectrum, the quasi-static distribution function produced by the diffusion can be obtained by solving

$$\left[\left(1-\frac{k_\parallel v_\parallel}{\omega}\right)\frac{\partial}{\partial v_\perp} + \frac{k_\parallel v_\perp}{\omega}\frac{\partial}{\partial v_\parallel}\right]\bar{f}(v_\perp, v_\parallel) = 0. \quad (2.152)$$

The general solution of the above equation is

$$\begin{aligned}\bar{f}(v_\perp, v_\parallel) &= \bar{f}\left(w = \frac{v_\perp^2}{2} + \frac{v_\parallel^2}{2} - \frac{\omega}{k_\parallel}v_\parallel\right), \\ v_\parallel &= \frac{\omega-\omega_c}{k_\parallel}.\end{aligned} \quad (2.153)$$

When the distribution function diffuses to the above form, we can show that the growth (damping) rate vanishes, consequently the above distribution function is quasi-static. From Eq. (2.153), we see that the change in total energy E $(\propto v_\perp^2 + v_\parallel^2)$ of a particle associated with the change in the parallel energy $E_{||}(\propto v_{||}^2)$ is given by

$$dE/dE_{||} = \omega/k_{||}v_{||} = \omega/(\omega - \omega_c). \quad (2.154)$$

Consequently if $\omega \ll \omega_c$, little change in the total energy of the particle is expected by the cyclotron wave; only pitch angle $\alpha\,(\tan^{-1} v_\perp/v_{||})$ diffuses. In such a case Eq. (2.151) can be reduced to a pitch angle diffusion equation:

$$\frac{\partial f(v, \alpha)}{\partial t} = \frac{1}{\sin\alpha}\,\frac{\partial}{\partial\alpha}\left(D_\alpha \sin\alpha\,\frac{\partial f}{\partial\alpha}\right)$$

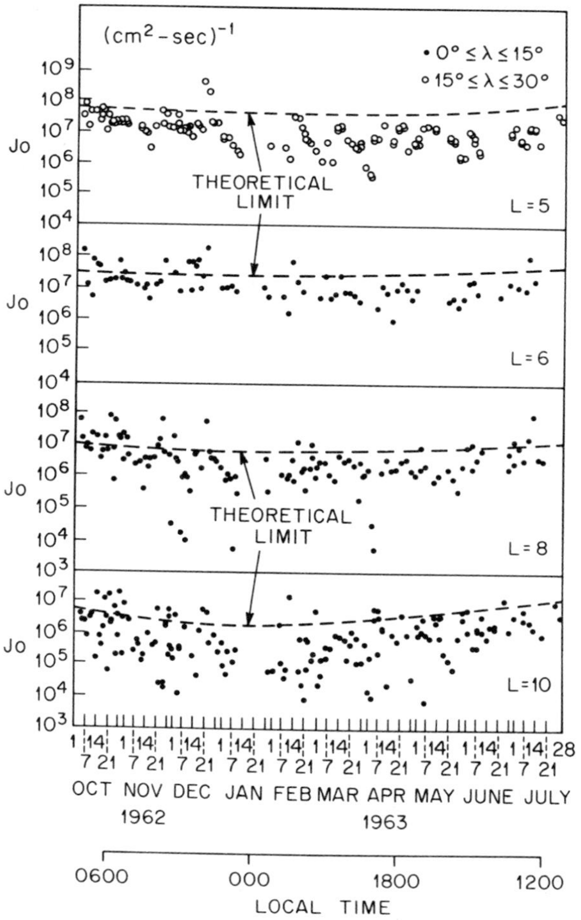

Fig. 22. Limitation of stably trapped electrons >40 keV as calculated by whistler wave turbulence. (After Kennel and Petschek, 1966)

where

$$D_\alpha = 2\pi^2 \omega_c^2 \sum_k \frac{1}{2} \left| \frac{B_k}{B} \right|^2 \frac{1}{|k_{||}|} \delta\left(v \cos\alpha - \frac{\omega - \omega_c}{k_{||}}\right) \tag{2.155}$$

and B_k is the flux density of the wave magnetic field.

Kennel and Petschek (1966) solved this velocity-space diffusion equation by assuming the noise amplitude of a whistler wave. The anisotropy of the pitch angle distribution gives the growth rate of the whistler wave carried by cold electrons. Using this growth rate and assuming a suitable reflection at the ionosphere, they calculated the magnitude of the whistler wave. By matching this calculated magnitude with the originally assumed amplitude, they closed the chain of quasi-linear equations and obtained the limit of stable trapping for electron flux (cf. Fig. 22) in the magnetosphere.

Let us now look at diffusion associated with the electrostatic waves. In a similar manner, the quasi-static distribution functions for this case can be obtained from Eq. (2.149)

$$\left(\frac{n\omega_c}{v_\perp} \frac{\partial}{\partial v_\perp} + k_{||} \frac{\partial}{\partial v_{||}}\right) \bar{f} = 0 \tag{2.156}$$

from which we have

$$\bar{f}(v_\perp, v_{||}) = \bar{f}\left(w = \frac{v_\perp^2}{2} - \frac{n\omega_c v_{||}}{k_{||}}\right), \quad v_{||} = \frac{\omega - n\omega_c}{k_{||}}. \tag{2.157}$$

Substituting the second expression, we see that the contour of constant w is an ellipsoid, instead of a sphere as in the previous case, given by

$$\frac{v_\perp^2}{2} + \left(v_{||} - \frac{\omega}{2k_{||}}\right)^2 = \frac{\omega^2}{4k_{||}^2} + w. \tag{2.158}$$

Consequently in this case both energy and pitch angle diffuse at nearly the same rate.

2.4 Hydromagnetic Instabilities

2.4a Introduction

In this section we will consider velocity space instabilities at very low frequencies ($\omega \ll \omega_{ci}$) as well as at very long wavelengths ($\lambda \gg \varrho_i = v_{Ti}/\omega_{ci}$). Instabilities in this regime, commonly called hydromagnetic instabilities, occur in a high β plasma and are caused by anisotropic pressure. One could treat these instabilities by using MHD equations with anisotropic pressure, but we will use kinetic equations here to show some significant differences from the results of the MHD equations. Consequently, let us first introduce the dielectric tensor obtained from the kinetic equation (Eq. (2.43)) for the hydromagnetic limit: $k_\perp v_{Ti}/\omega_{ci} \sim \omega/\omega_{ci} \ll 1$.

$$\begin{aligned}
\varepsilon_{xx} &= \sum_j \frac{\omega_{pj}^2}{\omega_{cj}^2}\left[\left\langle\left(1-\frac{k_\parallel v_{\parallel j}}{\omega}\right)^2\right\rangle - \frac{k_\parallel^2 \langle v_{\perp j}^2\rangle}{2\omega^2}\right] \\
\varepsilon_{xy} &= -\varepsilon_{yx} = i\sum_j \frac{\omega_{pj}^2}{\omega\omega_{cj}}\left(1-\frac{k_\parallel \langle v_{\parallel j}\rangle}{\omega}\right) \\
\varepsilon_{xz} &= \varepsilon_{zx} = \sum_j \frac{\omega_{pj}^2}{\omega_{cj}^2}\frac{k_\perp}{k_\parallel}\left[\frac{k_\parallel^2\langle v_{\perp j}^2\rangle}{2\omega^2} + \left\langle\frac{k_\parallel v_{\parallel j}}{\omega}\left(1-\frac{k_\parallel v_{\parallel j}}{\omega}\right)\right\rangle\right] \\
\varepsilon_{yy} &= \varepsilon_{xx} - \sum_j \frac{\omega_{pj}^2}{\omega_{cj}^2}\left(\frac{k_\perp^2\langle v_{\perp j}^2\rangle}{\omega^2} + \frac{k_\perp^2\langle v_{\perp j}^4\rangle}{4\omega^2} I_j\right) \\
\varepsilon_{yz} &= -\varepsilon_{zy} = -i\sum_j \frac{\omega_{pj}^2}{\omega\omega_{cj}}\left(\frac{k_\perp\langle v_{\parallel j}\rangle}{\omega} + \frac{k_\perp}{2k_\parallel}\langle v_{\perp j}^2\rangle I_j\right) \\
\varepsilon_{zz} &= -\sum_j \left[\frac{\omega_{pj}^2}{k_\parallel^2} I_j + \frac{\omega_{pj}^2}{\omega_{cj}^2}\left(\frac{k_\perp^2\langle v_{\perp j}^2\rangle}{2\omega^2} - \frac{k_\perp^2\langle v_{\parallel j}^2\rangle}{\omega^2}\right)\right]
\end{aligned} \qquad (2.159)$$

where

$$I_j = \int_{-\infty}^{\infty} dv_\parallel \int_0^\infty 2\pi v_\perp dv_\perp \frac{f_{0j}(v_\perp, v_\parallel)}{(v_\parallel - \omega/k_\parallel)^2}, \quad (\mathrm{Im}\,\omega > 0),$$

and

$$\langle v_j^n\rangle = \int_{-\infty}^{\infty} dv_\parallel \int_0^\infty 2\pi v_\perp dv_\perp v^n f_{0j}(v_\perp, v_\parallel).$$

In these expressions the effects of wave-particle resonance, anisotropic distribution, and the existence of relative drift between different species are retained to be used for a study of instabilities. j indicates different species of plasma particles and ω_{cj} is the cyclotron frequency that includes sign (i.e., for electrons $\omega_{cj} = -\omega_{ce} < 0$).

In the absence of a relative drift between electrons and protons, $\varepsilon_{xy} = \varepsilon_{yx} = 0$ and the wave equation (2.37) can be shown to reduce into two uncoupled equations (Kutsenko and Stepanov, 1960):

$$\frac{c^2 k_\parallel^2}{\omega^2} - \varepsilon_{xx} = 0 \qquad (2.160)$$

and

$$\frac{c^2 k^2}{\omega^2} - \left(\varepsilon_{yy} + \frac{\varepsilon_{yz}^2}{\varepsilon_{zz}} \right) = 0. \tag{2.161}$$

Note here that the wave-vector lies in xz plane, thus $k^2 = k_x^2 + k_z^2 \equiv k_\perp^2 + k_\|^2$. Eq. (2.160) represents an incompressional wave ($\boldsymbol{\nabla} \cdot \boldsymbol{v} = 0$, where $\boldsymbol{v}$ is the fluid velocity), while (2.161) represents a compressional wave ($\boldsymbol{\nabla} \cdot \boldsymbol{v} \neq 0$). For an isotropic plasma, the former represents (shear) Alfvén wave and the latter magnetosonic (or compressional Alfvén) wave and ion acoustic wave.

2.4b Mirror Instability

Let us first discuss the hydromagnetic instability generated by a large perpendicular pressure. This is an instability associated with the compressional mode, represented by Eq. (2.161). When, as in the magnetosphere, a small admixture of cold electrons (at least $\sim 10\%$ in density) is expected, this equation can be simplified because the parallel electric field, E_z, is short circuited by a large ε_{zz} and the ion acoustic wave can be decoupled from the compressional Alfvén wave. The resultant dispersion relation then becomes

$$\frac{c^2 k^2}{\omega^2} - \varepsilon_{yy} = 0 \tag{2.162}$$

where, if we assume a Maxwellian distribution in the parallel direction, ε_{yy} is given by

$$\varepsilon_{yy} = \sum_j \left\{ \frac{\omega_{pj}^2}{\omega_{cj}^2} - \frac{c^2}{\omega^2} \left[k_\|^2 \, \frac{\beta_{\perp j} - \beta_{\| j}}{2} + \right.\right.$$
$$\left.\left. + k_\perp^2 \beta_{\perp j} \left(1 + \frac{\beta_{\perp j}}{2\beta_{\| j}} Z'\left(\frac{\omega}{\sqrt{2}\, k_\| v_{T\| j}} \right) \right) \right] \right\}. \tag{2.163}$$

Z' is the derivative of the plasma dispersion function defined in Eq. (2.31) and $\beta_\perp$ and $\beta_\|$ are the perpendicular and parallel components of β given by

$$\beta_{\perp j} = \frac{1}{2} \frac{mn\langle v_{\perp j}^2 \rangle}{B_0^2/2\mu_0} = \frac{mnv_{T\perp j}^2}{B_0^2/2\mu_0} = 2\,\frac{\omega_{pj}^2 v_{T\perp j}^2}{\omega_{cj}^2 c^2}$$

and

$$\beta_{\| j} = \frac{mnv_{T\| j}^2}{B_0^2/2\mu_0} = 2\,\frac{\omega_{pj}^2 v_{T\| j}^2}{\omega_{cj}^2 c^2}.$$

The unstable root is found by taking a low phase velocity limit such that

$$\omega/k_{\|} v_{T\|i} \ll 1$$

and

$$\omega^2/k^2 v_A^2 \ll 1$$

in which case the dispersion relation of the compressional mode (2.162) reduces to

$$k_{\|}^2 \left(1+\sum_j \frac{\beta_{\perp j}-\beta_{\| j}}{2}\right)+k_{\perp}^2 \left[1+\sum_j \beta_{\perp j} \left(1-\frac{\beta_{\perp j}}{\beta_{\| j}}\right) - i\,\frac{\beta_{\perp i}^2}{\beta_{\| i}}\,\frac{\omega}{k_{\|} v_{T\| i}}\left(\frac{\pi}{2}\right)^{1/2}\right]=0 \tag{2.164}$$

where subscript i shows ions.

Solving for ω, we have

$$\omega = -i k_{\|} v_{T\| i}\left(\frac{2}{\pi}\right)^{1/2} \frac{\beta_{\| i}}{\beta_{\perp i}^2}\left[\frac{k_{\|}^2}{k_{\perp}^2}\left(1+\sum_j \frac{\beta_{\perp j}-\beta_{\| j}}{2}\right) + 1+\sum_j \beta_{\perp j}\left(1-\frac{\beta_{\perp j}}{\beta_{\| j}}\right)\right]. \tag{2.165}$$

One can immediately see from this expression that the instability ($\mathrm{Im}\,\omega > 0$) occurs either when

$$1+\sum_{\text{species}} \frac{\beta_{\perp}-\beta_{\|}}{2} < 0 \tag{2.166}$$

(for large $k_{\|}/k_{\perp}$; almost parallel propagation), or when

$$1+\sum_{\text{species}} \beta_{\perp}\left(1-\frac{\beta_{\perp}}{\beta_{\|}}\right) < 0 \tag{2.167}$$

(for small $k_{\|}/k_{\perp}$; almost perpendicular propagation).

Eq. (2.166) is the condition of the hose instability which will be discussed in the next subsection. Eq. (2.167) which is satisfied for $\beta_{\perp} > > \beta_{\|} \sim 1$, is a complementary situation to the hose instability. The instability represented by this condition is called the mirror instability (Rosenbluth, 1956; Chandrasekhar, 1961; Sagdeev, 1966).

Let us briefly discuss the physical implication of the mirror instability. As can be seen from the condition of instability, the mirror instability is not the instability of the compressional Alfvén wave whose dispersion relation is given by $\omega^2/k^2 v_A^2 \cong 1$. In fact, we can see from Eq. (2.165) that ω is purely imaginary, whether growing or decaying. The mirror instability is hence an instability of a non-oscillatory compressional mode. Such a non-oscillatory mode appears also as a drift wave when diamagnetic drift motion is considered. The Im $\omega < 0$ solution of Eq. (2.165) for a plasma with an isotropic pressure, $\beta_\perp = \beta_{||}$, represents transit-time damping (Stix, 1962). Transit-time damping is the magnetic analogue of Landau damping where $\mu B_{||}$ acts like an electrostatic potential φ, where μ is the magnetic moment. When a compressional wave is set up, it is

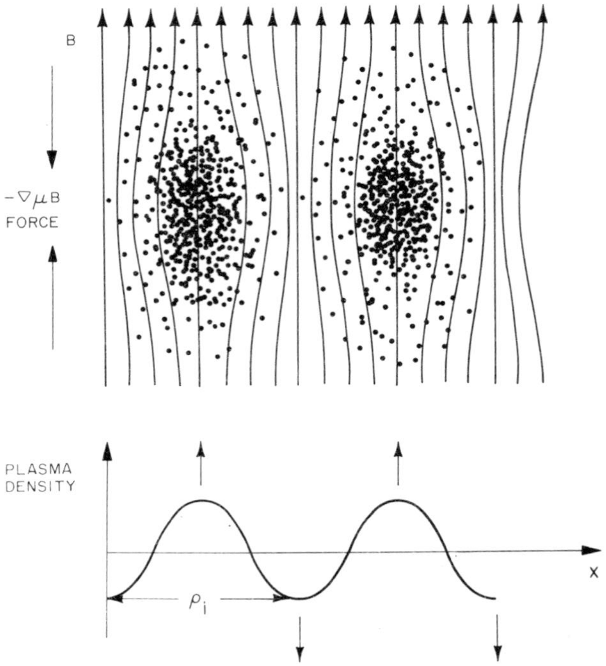

Fig. 23. Schematic diagram of the drift mirror instability. When the instability condition is satisfied, the $-\nabla\mu B$ force that accelerates plasma parallel to the field lines into the weaker field region, excludes the fields further by the diamagnetic effect and the process grows. This effect occurs over a region where the instability condition is satisfied. The growth rate becomes maximum when the spatial periodicity is roughly the proton gyroradius $\varrho_i(\sim 100$ km). In the presence of proton diamagnetic drift, such periodic bumps of plasma drift also with velocity v_d, hence for a stationary observer an oscillatory proton density, which is out of phase with respect to the field intensity, can be observed with the oscillation frequency given by v_d/ϱ_i

ordinarily damped out by the transit time damping. However, when $\beta_\perp > \beta_\parallel \sim 1$, the diamagnetic repulsion of the plasma, which is trapped in the local mirror field created by the wave, excludes the magnetic field. This accelerates the flow of plasma into the thus deepened well of the local mirror, and therefore the perturbation grows (cf. Fig. 23).

We present here an example of a possible occurrence of the mirror instability near the equatorial plane in the magnetosphere during a geomagnetic storm on April 18, 1965. Fig. 24 shows the variations of the magnetic field and proton fluxes published by Brown *et al.* (1968). The observation was made by Explorer 26 which was located at 5.11 R_E and a latitude of 17° at 1400 LT. In this figure can be seen (a) a strong diamagnetic effect (indicating $\beta \sim 1$) starting at 6:15 and suddenly terminating at 6:20 shown by point A, (b) a large anisotropy of the proton fluxes (θ is measured with respect to the direction of the magnetic field) indicating $\beta_\perp > \beta_\parallel$, and (c) the start of large out of phase oscillations of the field and the fluxes shortly after the point A.

The first two facts strongly indicate an occurrence of the mirror instability. When such an instability is set up, it can terminate a further increase and anisotropization of the proton flux and can trigger succeeding oscillations. In fact, it was found corresponding to the point A, $\beta_\perp/\beta_\parallel \sim 2$ and $\beta_\perp \sim 1$ satisfying the condition of instability (Hasegawa, 1969; Lanzerotti *et al.*, 1969).

Because the mode causing the mirror instability is non-oscillatory, other effects are required to explain the oscillations clearly observed

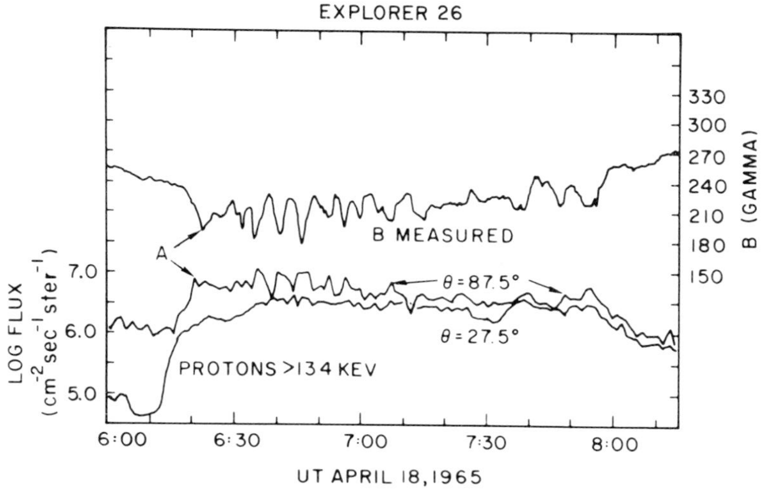

Fig. 24. Observed variations of proton fluxes and the geomagnetic field at $\sim 5.11\ R_E$. (After Brown *et al.*, 1968)

after 6:30. The oscillation may be produced by coupling with the drift wave created by the ion drift perpendicular to the magnetic field. This would give rise to a real frequency $\omega \sim k_\perp v_d$, where v_d is the proton diamagnetic drift speed and $k_\perp$ is the perpendicular wave number, which may be chosen to be $\omega_{ci}/\langle v_\perp \rangle_i$, corresponding to its value for a maximum growth rate. Observed values of average proton energy (~ 20 keV) and the magnetic field ($\sim 200\ \gamma$) give a frequency which is in good agreement with the observed frequency.

The phase relation between the proton flux and the magnetic field is another important point to check. Under the MHD approximation, only the sound wave has the observed out of phase relation. However, because $T_e < T_i$ this mode is heavily damped in the magnetosphere. On the other hand, if the kinetic equation is used allowing anisotropic pressure, one can show that the phase relation between the proton density perturbation n_1 and the magnetic field perturbation B_1 for the compressional modes becomes (Hasegawa, 1969).

$$\frac{n_1}{n_0} \sim \left(1 - \frac{\beta_{\perp i}}{\beta_{\| i}}\right) \frac{B_1}{B_0}.$$

Therefore, if $\beta_\perp > \beta_\|$, n_1 and B_1 in fact become out of phase. This is because, when $\beta_\perp > \beta_\|$, the effect of $-\nabla \mu B$ force which accelerates plasma into the lower flux region parallel to B_0 overbalances the effect of $\boldsymbol{E} \times \boldsymbol{B}$ drift which moves plasma with the magnetic field in the perpendicular direction.

2.4c Fire Hose Instability

We have seen in Eq. (2.166) that the non-oscillatory compressional mode becomes unstable for a large value of $k_\|/k_\perp$, or for almost parallel propagation, when $\beta_\| > \beta_\perp \sim 0(1)$. We now show that the same instability condition applies to the incompressional (shear) Alfvén wave. If we substitute ε_{xx} of Eq. (2.159) into the dispersion relation of the incompressional mode (2.160), we have

$$\frac{\omega^2}{k_\|^2 v_A^2} = 1 - \frac{1}{2} \sum_j (\beta_{\| j} - \beta_{\perp j}). \tag{2.168}$$

Consequently, we see that the same instability condition, $\beta_\| - \beta_\perp > 2$ applies to the shear Alfvén wave. Because the physical mechanism of this instability is similar to that which generates oscillations in a water hose when the water pressure exceeds a critical value, this is called fire hose instability (Rosenbluth, 1956; Parker, 1958).

Parker (1958) suggested that the hose instability might explain the relatively isotropic velocity distribution of solar wind. The solar wind expands along the solar magnetic field. Because the expansion is a slow process, the magnetic moment $mv_\perp^2/2B_0(r)$ is conserved, thus $v_\perp$ will decrease as B_0 decreases away from the sun, causing a large anisotropic distribution of the type $v_{T\perp} \ll v_{T\|}$. In addition the pressure ratio of plasma to the magnetic field in the solar wind, β, is of order unity, thus it is quite natural that the hose instability may be responsible for the rather isotropic distribution observed near the earth ($v_{T\|} \sim \sqrt{2}\, v_{T\perp}$). Further study, however, seems to indicate that collisional isotropization can explain some of the observed results, and if an instability is responsible, it may be an instability of the compressional mode caused by a combined effect of anisotropic β ($\beta_\| > \beta_\perp$) and a difference of heat fluxes between electrons and protons in the parallel direction [the peaks of the velocity distributions of protons and electrons in the parallel (to the solar magnetic field) direction are shifted even if their average velocity in this direction is the same] (Forslund, 1970; Schulz and Eviatar, 1971).

2.5 Instabilities in Partially Ionized Plasmas

2.5a Introduction

When a plasma is only partially ionized, collisions between charged and neutral particles become important. Such collisions allow the existence of a *dc* electric field in directions both parallel and perpendicular to the magnetic field. Such an electric field induces current in the direction parallel to the electric field (Pederson current) as well as perpendicular to the electric and the magnetic field (Hall current). The plasma in such a case is also generally nonuniform. The nonuniformity and the *dc* electric field produce electrostatic instabilities which are unique in the partially ionized plasma. In this section, we will discuss the general properties of such instabilities (Subsection 2.5b) and their application to the ionosphere (Subsection 2.5c).

2.5b General Properties of Instabilities

We consider a model plasma, which is imbedded in uniform magnetic and electric fields $\boldsymbol{B}_0$ and $\boldsymbol{E}_0$. The plasma has its density gradient in the direction transverse to the ambient magnetic field. The coordinate system we take here is shown in Fig. 25. The transverse electric field E_{0x}, in the direction of the density gradient, is either an ambipolar field or that plus

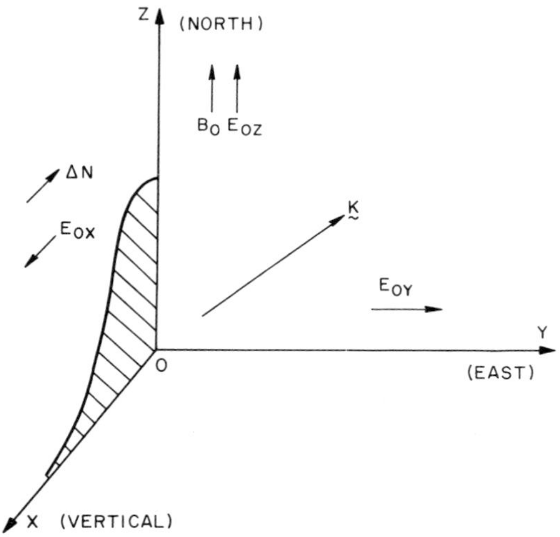

Fig. 25. Coordinate system used in the derivation of the instability conditions in Subsection 2.5. In the equatorial region, x, y, z correspond to vertical (upward), east and north directions, respectively

a field from an external source. We consider here a longitudinal electrostatic wave propagating in an arbitrary direction in this plasma.

First, let us study the behavior of electrons. The density gradient and the drift may contribute to create an active conductivity ($\mathrm{Re}\,\sigma<0$). The necessary equations are the equation of motion

$$n\boldsymbol{v} = -\mu_e n(\boldsymbol{E}+\boldsymbol{v}\times\boldsymbol{B}) - D_e \nabla n \tag{2.169}$$

and the equation of continuity

$$\frac{\partial n}{\partial t} + \nabla\cdot(n\boldsymbol{v}) = -\delta n \tag{2.170}$$

where μ_e is the electron mobility ($=e/\nu_e m_e$), D_e is the electron diffusion constant ($=v_{Te}^2/\nu_e$), ν_e is the electron-neutral collision rate, v_{Te} is the electron thermal speed, and δ is the recombination frequency, which we assume to be negligibly small. The inertia term $\dot{\boldsymbol{v}}$ is ignored in Eq. (2.169) because we are considering a low-frequency range where $\omega \ll \omega_{ce}$.

The unperturbed quantities, for which we use subscript 0, are obtained from Eqs. (2.169) and (2.170)

$$v_{0z} = -\mu_e E_{0z}, \tag{2.171}$$

$$v_{0x} = \frac{\kappa D_e - \mu_e E_{0x} + \mu_e^2 B_0 E_{0y}}{1 + \mu_e^2 B_0^2}, \tag{2.172}$$

$$v_{0y} = \frac{\mu_e B_0 (\kappa D_e - \mu_e E_{0x}) - \mu_e E_{0y}}{1 + \mu_e^2 B_0^2} \tag{2.173}$$

where κ is the magnitude of the density gradient

$$\kappa = -\frac{d(\ln n_0)}{dx}. \tag{2.174}$$

For the perturbed quantities, we assume a phase factor of $\exp i(\boldsymbol{k} \cdot \boldsymbol{x} - \omega t)$, and use subscript 1. The assumption of a longitudinal disturbance allows us to use a scalar potential φ_1 for the electric field

$$\boldsymbol{E}_1 = -i\boldsymbol{k}\varphi_1 \tag{2.175}$$

Now we can solve Eqs. (2.169) and (2.170) for the perturbed density n_1 in terms of the perturbed potential φ_1. Substituting the result into the definition of longitudinal conductivity shown in Eq. (1.60) we can obtain the electron conductivity

$$\sigma_e = \frac{\omega \varepsilon_0 \omega_{pe}^2 [ik_0^2 + \kappa(\mu_e B_0 k_y - k_x)/(1 + \mu_e^2 B_0^2)]}{k^2 [i\nu_e(\omega - \boldsymbol{k} \cdot \boldsymbol{v}_E) - v_{Te}^2 k_0^2]} \tag{2.176}$$

$\boldsymbol{v}_E$ is the unperturbed velocity due to the electric field, namely

$$\boldsymbol{v}_E = -\mu_e E_{0z} \boldsymbol{e}_z - \frac{\mu_e E_{0x} - \mu_e^2 B_0 E_{0y}}{1 + \mu_e^2 B_0^2.} \boldsymbol{e}_x - \frac{\mu_e E_{0y} + \mu_e^2 B_0 E_{0x}}{1 + \mu_e^2 B_0^2} \boldsymbol{e}_y \tag{2.177}$$

and

$$k_0^2 = k_z^2 + \frac{k_x^2 + k_y^2}{1 + \mu_e^2 B_0^2}. \tag{2.178}$$

The active range of σ_e can be immediately obtained from the condition that $\mathrm{Re}\,\sigma_e < 0$, and is given by

$$0 < \omega < \boldsymbol{k} \cdot \boldsymbol{v}_E + \omega_e^* \tag{2.179}$$

where

$$\omega_e^* = \kappa v_T^2 k_y / \omega_{ce} \tag{2.180}$$

is the electron drift wave frequency.

Thus, from Eq. (2.179), we can see that both the electric field drift *and* the diamagnetic drift (due to the density gradient) contribute to make σ_e active. If the sign of ω_e^* and $\boldsymbol{k} \cdot \boldsymbol{v}_E$ is the same, these two effects cooperate to produce the active conductivity. In such a case, it can be shown that the electron dielectric constant possesses a property similar to that of a negative energy wave (Hasegawa, 1968), and an instability can be generated by a coupling with even a purely resistive medium or collision-dominated ions. ω_e^* and $\boldsymbol{k} \cdot \boldsymbol{v}_E$ have the same sign when the direction of electron drift due to the *dc* electric field $\boldsymbol{E}_0$ and that of the electron diamagnetic drift coincide. For example, for a wave propagating in the positive y direction, ω_e^* is positive when the density decreases in the positive x direction ($\kappa > 0$, $k_y > 0$) and $\boldsymbol{k} \cdot \boldsymbol{v}_E$ becomes positive when $v_{Ey} > 0$ or when $E_{0y} < 0$ and/or $E_{0x} < 0$. In other words, if the direction of the *dc* electric field is in the direction of the nonuniformity, $\boldsymbol{E}_0 \cdot \nabla n_0 > 0$, ω_e^* and $\boldsymbol{k} \cdot \boldsymbol{v}_E$ have the same sign and the instability is more easily excited; if $\boldsymbol{E}_0 \cdot \nabla n_0 < 0$, the effects of ∇n_0 and $\boldsymbol{E}_0$ mutually stabilize each other. For example, if we take the ambipolar electric field, which at the limit of $T_i = 0$ is given by

$$E_{0x} = \frac{\omega_e^* B_0}{k_y (1 + \mu_i \mu_e B_0^2)} > 0 \tag{2.181}$$

the y component of $\boldsymbol{v}_E$ becomes, when $\mu_e B_0 \gg 1$,

$$v_{Ey} = -\frac{E_{0x}}{B_0} < 0 . \tag{2.182}$$

Therefore, the ambipolar electric field, which is directed along the positive x axis (i.e., $\boldsymbol{E}_0 \cdot \nabla n_0 < 0$), has a stabilization effect (reducing the active nature of σ, produced by ∇n_0). However, an externally applied electric field that is directed along the negative x axis ($\boldsymbol{E}_0 \cdot \nabla n_0 > 0$) enhances the instability (Simon, 1963). On the other hand, any electric field in the z direction enhances the instability for a wave with $\boldsymbol{k}$ directed such that $\boldsymbol{k} \cdot \boldsymbol{E} < 0$.

To demonstrate the importance of the cooperative effect of the density gradient and the *dc* electric field, let us examine a simple case in which ions are collision dominated by choosing a frequency range such that $\omega \ll \omega_{ci} \ll v_i$, where v_i is the ion-neutral collision frequency.

For such a case, the ions constitute a simple resistive medium whose conductivity is given by

$$\sigma_i = \varepsilon_0 \omega_{pi}^2 / v_i \tag{2.183}$$

where ω_{pi} is the ion plasma frequency. The condition for the instability is given by Eqs. (2.176) and (2.183) as $-\mathrm{Re}(\sigma_e) > \varepsilon_0 \omega_{pi}^2/v_i$, at $\mathrm{Im}(\sigma_e) = 0$, for $\omega \neq 0$; or, explicitly,

$$\frac{\omega_{pe}^2 k_0^2}{v_e k^2}\left(\frac{\omega_e^* \boldsymbol{k} \cdot \boldsymbol{v}_E}{k_0^4 D_e^2} - 1\right) - \frac{\omega_{pi}^2}{v_i} > 0. \qquad (2.184)$$

When E_{0x} is the ambipolar field, Eq. (2.184) reduces to the condition of the collisional helical instability as derived by Kadomtsev (1965). Because the ambipolar electric field has a stabilizing effect ($k_y v_{Ey} < 0$), the instability condition, Eq. (2.184), is satisfied only in the presence of E_{0z}. However, an instability is possible even in the absence of E_{0z} if a negative E_{0x} is applied externally. The fact that a *dc* electric field in a partially ionized plasma produces an instability is because the $\boldsymbol{E}_0 \times \boldsymbol{B}_0$ drift speed of electrons and ions differs because of the difference in the value of $\mu B (= \omega_c/v)$, the ratio of the cyclotron frequency to the collision frequency to neutrals, and produces a "two stream" effect.

2.5c Application to Ionospheric Plasma Instabilities

During the period of 1960 to 1963, Bowles *et al.* (1963) performed a series of radar scattering experiments from ionosphere at Jicamarca station near the equator and showed that the equatorial electroject at an altitude of ~ 100 km is imbedded with large density irregularities with wavelengths of a few meters.

Later, in 1967, Cohen and Bowles (1967) reported weaker but more commonly observed irregularities that move with a speed slower than the former (which moves at the ion acoustic speed) and have longer wavelengths (tens to hundreds of meters). The former irregularity is generally called type I and the latter type II.

Immediately after the report of Bowles *et al.* (1963), Farley (1963), Buneman (1963) and Maeda *et al.* (1963) have reported theories of possible plasma instabilities that attempt to explain the observed irregularities.

Buneman-Farley's theory is essentially based on the two stream instability of the electron Hall current ($\boldsymbol{k} \cdot \boldsymbol{v}_E$ effect in Eq. (2.179)), while the theory of Maeda *et al.* is based on the drift wave instability (ω_e^* effect in Eq. (2.179)). Maeda *et al.*, however, considered the electric field only in the direction of electron drift (E_{0y} in Fig. 25) and thus the contribution of $\boldsymbol{k} \cdot \boldsymbol{v}_E$ was smaller than in the perpendicular direction (E_{0x}) by a factor of $(\mu_e B_0)^{-1}$ $(= v_e/\omega_{ce} \ll 1)$. Tsuda *et al.* (1966) corrected this point and introduced the Hall electric field. In this way ω_e^* *and*

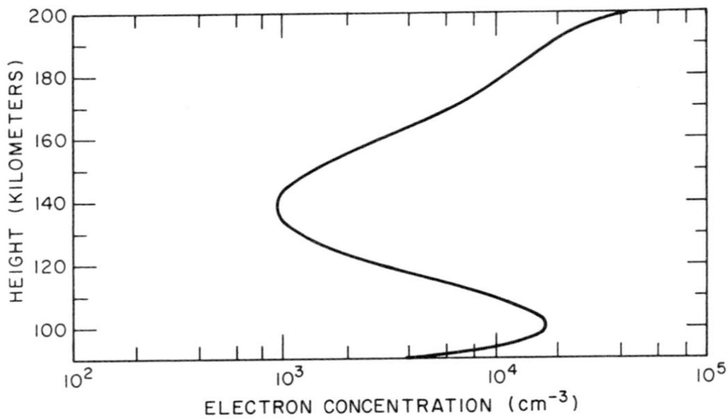

Fig. 26. Typical electron density profile as a function of altitude. (After Reid, 1968)

$\boldsymbol{k} \cdot \boldsymbol{v}_E$ contributed to the instability (Simon instability) and a much lower threshold in $\boldsymbol{E}_0$ was found than those used by Farley and Buneman. In addition, as was pointed out in Subsection 2.5b, because of the co-operative effect of ω_e^* and $\boldsymbol{k} \cdot \boldsymbol{v}_E$, this instability occurs at much lower frequencies ($\omega \ll v_i$) and hence at much longer wavelengths, than the Farley-Buneman two stream instability. However, as was mentioned in the previous subsection, a cooperative effect exists between ω_e^* and $\boldsymbol{k} \cdot \boldsymbol{v}_e$ only when $\boldsymbol{E}_0 \cdot \nabla n_0$ is positive. Thus, $\boldsymbol{E}_0$ must be directed in the same direction as ∇n_0, hence must be vertical. Because ∇n_0 in the vertical direction changes its sign (Fig. 26) in the ionosphere, if the direction of $\boldsymbol{E}_0$ is given, there are favorable and unfavorable altitudes for the occurrence of the Simon instability; the instability conditions become a strong function of altitude (Reid, 1968).

Let us now derive the specific condition of instability for the ionospheric parameters: $v_e \sim 4 \times 10^4$ sec^{-1}, $v_i \sim 2.5 \times 10^3$ sec^{-1}, $\omega_{ce} \sim 5 \times 10^6$ sec^{-1}, $\omega_{ci} \sim 10^2$ sec^{-1}, the ion acoustic speed $[(T_e + T_i)/m_i]^{1/2} \sim 3 \times 10^2$ m/sec, and $\kappa \sim 10^{-4}$ m^{-1}. We note then v_e/ω_{ce} $(=(\mu_e B_0)^{-1}) \sim 10^{-2}$ while ω_{ci}/v_i $(=(\mu_i B_0)) \sim 10^{-1}$ to 10^{-2}. Hence, if we ignore the parallel electric field E_{0z}, $\boldsymbol{k} \cdot \boldsymbol{v}_E$ becomes, from Eq. (2.177),

$$\boldsymbol{k} \cdot \boldsymbol{v}_E \sim -\frac{k_y E_{0x}}{B_0}. \tag{2.185}$$

For ions, $\boldsymbol{v}_E$ is mostly directed in the x direction; thus for a wave propagating mostly in the y direction, $\boldsymbol{k} \cdot \boldsymbol{v}_E$ is negligible. If we intend to cover a frequency range ω such that the ion inertia is important, we must retain the $d\boldsymbol{v}/dt$ term in the equation of motion (2.169). The effect of a density gradient for ions is negligible because for $k_\perp \sim k_y$, and $\mu_i B_0 \ll 1$, hence the term multiplied by κ in Eq. (2.176) as expressed for ions vanishes. Consequently the ion conductivity becomes

$$\sigma_i = \frac{i\omega\varepsilon_0\omega_{pi}^2}{\omega(i\nu_i+\omega) - v_{Ti}^2 k^2}. \tag{2.186}$$

The dispersion relation is obtained from $\sigma_i + \sigma_e = 0$ by substituting (2.176) and (2.186),

$$(\omega - \boldsymbol{k}\cdot\boldsymbol{v}_E) = \frac{i}{\mu_e\mu_i B_0^2}\left[\omega\left(\frac{\omega}{\nu_i}+i\right) - \frac{k_y^2 c_s^2}{\nu_i}\right]\left(1 - \frac{i\kappa\mu_e B_0}{k_y}\right). \tag{2.187}$$

We can obtain the frequency of the wave and the growth rate by putting $\omega = \omega_k + i\gamma_k$, $(\omega_k \gg \gamma)$ as

$$\omega_k = \frac{\boldsymbol{k}\cdot\boldsymbol{v}_E}{1+(\mu_e\mu_i B_0^2)^{-1}} \tag{2.188}$$

and

$$\gamma_k = \frac{1}{1+\mu_e\mu_i B_0^2}\left\{\frac{1}{\nu_i}\left(\omega_k^2 - k_y^2 c_s^2\right) + \omega_k \frac{\kappa\mu_e B_0}{k_y}\right\}. \tag{2.189}$$

The above expression was obtained by Rogister and D'Angelo (1970). As expected, the instability ($\gamma_k > 0$) occurs by the combined effect of $\boldsymbol{E}_0 \times \boldsymbol{B}_0$ electron drift and density gradient. We can see that an instability of an ion acoustic wave occurs when $\boldsymbol{k}\cdot\boldsymbol{v}_E > k_y c_s[1+(\mu_e\mu_i B_0^2)^{-1}]$ even in the absence of the density gradient ($\kappa = 0$); this corresponds to the Buneman-Farley instability. On the other hand, in the presence of a density gradient (that has $\nabla n_0 \cdot \boldsymbol{E}_0 > 0$), the third term in the bracket can contribute to make $\gamma_k > 0$ even if $\omega_k < k c_s$. The instability is thus produced at lower frequencies, hence at longer wavelengths. Because of this, it is generally considered that the type I irregularity is caused by Buneman-Farley type two stream instability, while the type II irregularity is generated by a Simon type instability as was shown first by Tsuda *et al.* (1966).

However, Sudan *et al.* (1973) have shown that the type I irregularity can be generated by nonlinear development of the type II irregularity, which as has been shown here, is more easily excited. This theory can explain the generation of type I irregularities even with a relatively weak electric field. In fact, if we go back to Eq. (2.187), and solve it for an instability condition for the case of zero dc electric field E_0, we obtain

$$\frac{\omega_e^*}{1+(\mu_e \mu_i B_0^2)^{-1}} > k_y c_s, \tag{2.190}$$

which is the condition obtained by Moiseev and Sagdeev (1963).

We note finally that the Simon instability has also been used to explain the formation of a striation in an artificial plasma cloud injected in the ionosphere or magnetosphere (for examples: Linson and Workman, 1970; Simon, 1970).

References for Chapter 2

Akasofu, S. I.: Magnetospheric substorm as a discharge process. Nature **221**, 1020 (1969)
Berk, H. L., Book, D. L.: Plasma wave regeneration in inhomogeneous media. Phys. Fluids **12**, 649 (1969)
Bernstein, I. B.: Waves in a plasma in a magnetic field. Phys. Rev. **109**, 10 (1958)
Bernstein, I. B., Dawson, J. M.: Hydromagnetic instabilities caused by runaway electrons. Paper presented at Controlled Thermonuclear Conference, Washington, D. C. (1958)
Bowles, K. L., Balsley, B. B., Cohen, R.: Field aligned E region irregularities identified with acoustic plasma waves. J. Geophys. Res. **68**, 2485 (1963)
Brice, N.: An explanation of triggered VLF emissions. J. Geophys. Res. **68**, 4626 (1963)
Briggs, R. J.: Electron stream interaction with plasmas, Cambridge, Mass.: MIT, 1964
Brown, W. L., Cahill, L. J., Davis, L. R., McIlwain, C. E., Roberts, C. S.: Acceleration of trapped particles during magnetic substorm on April 18, 1965. J. Geophys. Res. **73**, 153 (1968)
Buneman, O.: Instability, turbulence, and conductivity in current carrying plasma. Phys. Rev. Letters **1**, 8 (1958)
Buneman, O.: Excitation of field aligned sound waves by electron streams. Phys. Rev. Letters **10**, 285 (1963)
Byers, J. A., Grewel, M.: Perpendicular propagating plasma cyclotron instabilities simulated with a one dimensional computer model. Phys. Fluids **13**, 1819 (1970)
Chandrasekhar, S.: Hydrodynamic and hydromagnetic stability, Chap. 13, Oxford; Clarendon, 1961
Cloutier, P. A., Anderson, H. R., Park, R. J., Vondrak, R. R., Spiger, R. J., Sandel, B. R.: Detection of geomagnetically aligned currents associated with an auroral arc. J. Geophys. Res. **75**, 2595 (1970)
Cohen, R., Bowles, K. L.: Secondary irregularities in the equatorial electrojet. J. Geophys. Res. **72**, 885 (1967).
Cornwall, J. M., Coroniti, F. V., Thorne, R. M.: Turbulent loss of ring current protons. J. Geophys. Res. **75**, 4699 (1970)
Criswell, D. R.: Pc 1 micropulsation activity and magnetospheric amplification of 0.2- to 5.0-Hz hydromagnetic waves. J. Geophys. Res. **74**, 205 (1969)
Cummings, W. D., Dessler, A. J.: Field-aligned currents in the magnetosphere. J. Geophys. Res. **72**, 1007 (1967).
Dawson, I.: On Landau damping. Phys. Fluids **4**, 869 (1961)
DeForest, S. E., McIlwain, C. E.: Plasma clouds in the magnetosphere. J. Geophys. Res. **76**, 3587 (1971).

Dory, R. A., Guest, G. E., Harris, E. G.: Unstable electrostatic plasma waves propagating perpendicular to a magnetic field. Phys. Rev. Letters **14**, 131 (1965)

Drummond, W. E., Rosenbluth, M. N.: Anomalous diffusion arising from microinstabilities in a plasma. Phys. Fluids **5**, 1507 (1962)

Eviatar, A., Wolf, R. A.: Transfer processes in the magnetosphere. J. Geophys. Res. **73**, 5161 (1968)

Farley, D. T., Jr.: A plasma instability resulting in field-aligned irregularities in the ionosphere. J. Geophys. Res. **68**, 6083 (1963)

Forslund, D. W.: Instabilities associated with heat conduction in solar wind and their consequences. J. Geophys. Res. **75**, 17 (1970)

Forslund, D. W., Morse, R. L., Nielsen, C. W.: Electron cyclotron drift instability. Phys. Rev. Letters **25**, 1266 (1970)

Fredericks, R. W., Crook, G. M., Kennel, C. F., Green, I. M., Scarf, F. L.: OGO 5 observations of electrostatic turbulence in bow shock magnetic structures. J. Geophys. Res. **75**, 3751 (1970)

Fried, B. D., Conte, S. E.: The plasma dispersion function. New York: Academic Press, 1961

Fried, B. D., Gould, R.: Longitudinal ion oscillations in a hot plasma. Phys. Fluids **4**, 139 (1961)

Fukunishi, H.: Occurrence of sweepers in the evening sector following the onset of magnetosphere substorms. Rept. of Ionosphere and Space Res. Japan **23**, 21 (1969)

Gintsburg, M. A.: The generation of plasma waves by solar corpuscular streams. Astronomicheskii Zh. **37**, 979 (1960) [English Transl. Soviet Astronomy – AJ **4**, 913 (1961)]

Gitomer, S. J., Forslund, D. W., Rudsinski, L.: Numerical simulation of the Harris instability in two dimensions. Phys. Fluids **15**, 1570 (1972)

Gruber, S., Klein, M. W., Auer, P. L.: High frequency velocity space instabilities. Phys. Fluids **8**, 1504 (1965)

Gurnett, D. A., Frank, L. A.: ELF noise bands associated with auroral electron precipitation. J. Geophys. Res. **77**, 3411 (1972).

Hall, L. S., Heckrotte, W., Kamash, T.: Ion cyclotron electrostatic instabilities. Phys. Rev. **139**, A117 (1965).

Hamberger, S. M., Jancarik, J.: Dependence of "anomalous" conductivity of plasma on the turbulent spectrum. Phys. Rev. Letters **25**, 999 (1970).

Harris, E. G.: Plasma instabilities associated with anisotropic velocity distributions. J. Nucl. Energy, Part C, Plasma Phys. **2**, 138 (1961)

Hasegawa, A.: Microinstabilities in transversely magnetized semiconductor plasmas. J. Appl. Phys. **36**, 3590 (1965)

Hasegawa, A.: Theory of longitudinal plasma instabilities. Phys. Rev. **169**, 204 (1968)

Hasegawa, A.: Drift mirror instability in the magnetosphere. Phys. Fluids **12**, 2642 (1969)

Hasegawa, A.: Excitation and propagation of an up streaming electromagnetic wave in the solar wind. J. Geophys. Res. **77**, 84 (1972)

Hasegawa, A., Birdsall, C. K.: Sheet-current plasma model for ion-cyclotron waves. Phys. Fluids **7**, 1590 (1964)

Hruska, A.: Cyclotron instabilities in the magnetosphere, J. Geophys. Res. **71**, 1377 (1966)

Jackson, J. D.: Longitudinal plasma oscillations. J. Nucl. Energy, Part C, Plasma Phys. **1**, 171 (1960)

Jacobs, J. A., Higuchi, Y.: Cyclotron amplification of geomagnetic micropulsations PC 1 in the magnetosphere. Planetary Space Sci. **17**, 2009 (1969)

Kadomtsev, B. B.: Plasma turbulence, p. 13(a); 89(b). New York, Academic Press, 1965

Kennel, C. F., Petschek, H. E.: Limit of stably trapped particle fluxes. J. Geophys. Res. **71**, 1 (1966)

Kennel, C.F., Scarf, F.L., Fredericks, R.W., McGehee, J.H., Coroniti, F.V.: VLF electric field observations in the magnetosphere. J. Geophys. Res. **75**, 6136 (1970)

Kimura, I.: Amplification of the VLF electromagnetic waves by a proton beam through the exosphere. An origin of the VLF emissions. Rept. Ionosphere Space Res. Japan **15**, 171 (1961)

Kimura, I., Matsumoto, H.: Hydromagnetic wave instabilities in a non-neutral plasma beam system. Radio Sci. **3**, 333 (1968)

Kindel, J.M., Kennel, C.F.: Topside current instabilities. J. Geophys. Res. **76**, 3055 (1971)

Krall, N.A., Liewer, P.C.: Turbulent heating and resistivity in cool electron Θ pinches. Phys. Fluids **15**, 1166 (1972)

Kutsenko, A.B., Stepanov, K.N.: Instability of plasma with anisotropic distributions of ion and electron velocities. Soviet Phys. JETP English Transl. **11**, 1323 (1960)

Landau, L.D.: On the vibrations of the electronic plasma. J. Phys. (USSR) **10**, 25 (1946)

Lanzerotti, L.J., Hasegawa, A., Maclennan, C.G.: Drift mirror instability in the magnetosphere, particle and field oscillation and electron heating. J. Geophys. Res. **74**, 5565 (1969)

Liemohn, H.B.: The cyclotron resonance amplification of whistlers in the magnetosphere. Boeing Sci. Res. Lab., Document D 1-82-0713 (1968)

Linson, L.M., Workman, J.B.: Formation of striations in ionospheric plasma clouds. J. Geophys. Res. **75**, 3211 (1970)

Maeda, K., Tsuda, T., Maeda, H.: Theoretical interpretation of the equatorial sporadic E layers. Phys. Rev. Letters **11**, 406 (1963).

Moiseev, S.S., Sagdeev, R.Z.: On the Bohm diffusion coefficient. Soviet Phys. JETP English Transl. **17**, 515 (1963)

Momata, H.: Stability of a uniform plasma composed of streams in the absence of an external field. Progr. Theoret. Phys. (Kyoto) **35**, 380 (1966)

Nishida, A.: Theory of irregular micropulsations associated with a magnetic bay. J. Geophys. Res. **69**, 947 (1964)

Nishihara, K., Hasegawa, A., Maclennan, C.G., Lanzerotti, L.J.: Elektrostatic instability plasmas. Phys. Rev. Letters **28**, 424 (1972)

Nishihara, K., Hasegawa, A., Maclennan, C.G., Lanzerotti, L.J.: Electrostatic instability exicted by an electron beam trapped in the magnetic mirror of the magnetosphere. Planetary Space Sci. **20**, 747 (1972)

Okuda, H., Hasegawa, A.: Computer experiments on plasma instabilities due to anisotropic velocity distributions. Phys. Fluids **12**, 676 (1969)

Parker, E.N.: Dynamic instability of an anisotropic ionized gas of low density. Phys. Rev. **109**, 1874 (1958)

Parker, E.N.: Small-scale nonequilibrium of the magnetopause and its consequences. J. Geophys. Res. **72**, 4365 (1967)

Pierce, J.R.: Possible fluctuations in electron streams due to ions. J. Appl. Phys. **19**, 231 (1948)

Post, R.F., Rosenbluth, M.N.: Electrostatic instabilities in finite mirror-confined plasmas. Phys. Fluids **9**, 730 (1966)

Reid, G.C.: The formation of small scale irregularities in the ionosphere. J. Geophys. Res. **73**, 1627 (1968)

Rogister, A., D'Angelo, N.: Type II irregularities in the equatorial electrojet. J. Geophys. Res. **75**, 3879 (1970)

Rosenbluth, M.N.: Los Alamos Scientific Laboratory Report LA-2030 (1956)

Russell, C.T., Childers, D.D., Coleman, P.J., Jr.: OGO-5 observations of upstream waves in the interplanetary medium: discrete wave packet. J. Geophys. Res. **76**, 845 (1971)

Sagdeev, R. Z.: Review of plasma physics, Vol. 4 (ed. by M. A. Leontovich), p. 32. New York: Consultants Bureau 1966.

Scarf, F. L., Bernstein, W., Fredericks, R. W.: Electron acceleration and plasma instabilities in the transition region. J. Geophys. Res. **70**, 9 (1965).

Scarf, F. L., Fredericks, R. W., Russell, C. T., Kivelson, M., Neugebayer, M., Chappell, C. R.: Observation of a current driven plasma instability at the outer-zone plasma sheet boundary. J. Geophys. Res. **78**, 2150 (1973)

Schulz, M., Eviatar, A.: Electron-temperature asymmetry and structure of the solar wind. Cosmic Electrodyn. **2**, 402 (1972)

Simon, A.: Instability of a partially ionized plasma in crossed electric and magnetic field. Phys. Fluids **6**, 382 (1963).

Simon, A.: Growth and stability of artificial ion cloud in the ionosphere. J. Geophys. Res. **75**, 6287 (1970)

Soper, G. K., Harris, E. G.: Effect of finite ion and electron temperature on the ion cyclotron resonant instability. Phys. Fluids **8**, 984 (1965)

Stix, T. H.: The theory of plasma waves, p. 196. New York: McGraw Hill, 1962

Sudan, R. N., Akinrimisi, J., Farley, D. T.: Generation of small scale irregularities in the equatorial electrojet. J. Geophys. Res. **78**, 240 (1973)

Swift, D. W.: A mechanism for energizing electrons in the magnetosphere. J. Geophys. Res. **70**, 3061 (1965).

Thorne, R. M.: Unducted whistler evidence for a secondary peak in the electron energy spectrum near 10 KeV. J. Geophys. Res. **73**, 4895 (1968)

Tsuda, T., Sato, T., Maeda, K.: Formation of sporadic E layers at temperate latitudes due to vertical gradients of charge density. Radio Sci. **1** (N. S.), 212 (1966).

Tsytovich, V. N.: Nonlinear Effects in Plasmas. (Trans. by M. Hamberger), p. 170. New York–London: Plenum Press, 1970

Weibel, E. S.: Spontaneously growing transverse waves in a plasma due to an anisotropic velocity distribution. Phys. Rev. Letters **2**, 83 (1959)

Young, T. S. T., Callen, J. D., McCune, J. E.: High frequency electrostatic waves in the magnetosphere. J. Geophys. Res. **78**, 1082 (1973)

Zmuda, A., Martin, J. H., Heuring, F. T.: Transverse magnetic disturbances at 1100 kilometers in the auroral region. J. Geophys. Res. **71**, 5033 (1966)

3. Macroinstabilities – Instabilities Due to Coordinate Space Nonequilibrium

3.1 Introduction

Nonuniformity in a plasma often provides a free energy source for plasma instabilities which work to smear out the original nonuniformity. Macroscopic structures in the plasma may change as a consequence of these instabilities, hence they are usually called macroscopic instabilities or macroinstabilities.

Plasmas can be nonuniform in various ways, but not all cases lead to instability. In many cases an instability is generated by the combined effect of a nonuniformity and the existence of a current or a plasma flow.

In Section 3.2 we introduce a genuine macroinstability, one that occurs only due to a plasma nonuniformity. The instability is called the drift wave or universal instability. In Section 3.3 and 3.4 we discuss plasma versions of classical fluid instabilities, the Rayleigh-Taylor instability and the Kelvin-Helmholtz instability. The last Section of this chapter, Section 3.5, will be devoted to instabilities of a nonuniform plasma with current.

3.2 Drift Wave Instabilities

The existence of a new mode called a drift wave, which arises due to a plasma nonuniformity, has been briefly introduced in the previous chapter. We study here how this mode may contribute to generate an instability. Depending on the value of β, the drift wave will couple to different modes. Consequently we treat problems in three ranges of values of β, low β ($\lesssim m_e/m_i$), medium β ($m_e/m_i \ll \beta \ll 1$), and high β (~ 1). The derivation, however, is limited to some special cases. For a more detailed treatment of drift wave instabilities, the reader should consult Krall (1968) or Mikhailovskii (1967).

3.2a Low β Case

The objective of this subsection is to introduce the basic mechanism of the drift wave instability. For this purpose, we assume a low temperature and low density plasma in which $\beta \lesssim m_e/m_i$, although it is inappropriate for most space plasmas. Under this assumption, however, we can treat the excited wave as electrostatic and the analysis is significantly simplified. We have already derived the equivalent conductivity for the electrostatic case in Eq. (2.139 a), taking into account the plasma nonuniformity

$$\sigma = \frac{-i\omega\varepsilon_0\omega_p^2}{k^2 v_T^2}\left[1+\frac{\omega-\omega^*}{\sqrt{2}\,k_{\|}v_T}\,Z\left(\frac{\omega}{\sqrt{2}\,k_{\|}v_T}\right)\right] \qquad (2.139\,\mathrm{b})$$

where $\omega^* = (-\kappa k_\perp v_T^2/\omega_c)$ is the drift wave frequency. Here we have ignored the contribution of polarization current (the last term of Eq. (2.139 a)). We have seen also that the conductivity becomes active, (Re $\sigma < 0$), when $\omega < \omega^*$. Let us derive the dispersion relation using this conductivity expressed for electrons and ions, and using also the quasi-neutrality condition $\sigma_i + \sigma_e = 0$. To obtain the drift wave we have to assume a parallel phase velocity range such that

$$v_{Te} \gg \frac{\omega}{k_{\|}} \gg v_{Ti}\,. \qquad (3.1)$$

then the dispersion relation becomes,

$$\frac{\omega_{pe}^2}{k^2 v_{Te}^2}\left[1+i\,\frac{\omega-\omega_e^*}{k_{\|}v_{Te}}\sqrt{\frac{\pi}{2}}\right] - \frac{\omega_{pi}^2}{k^2 v_{Ti}^2}\,\frac{\omega_i^*}{\omega} = 0\,. \qquad (3.2)$$

Here, both ω_e^* and ω_i^* are taken to be positive. If we note that $(\omega_{pe}^2/v_{Te}^2)/(\omega_{pi}^2/v_{Ti}^2) = (1/T_e)/(1/T_i)$, the above expression is satisfied by a real frequency $\omega = \omega_e^*$ and the imaginary part of ω vanishes at this frequency (which is the consequence of Re σ_e being zero at this frequency). The wave given by the dispersion relation $\omega = \omega_e^* = k_\perp v_{de}$ (v_{de} is the diamagnetic drift speed of electrons) is called a drift wave. The above result shows that the drift wave is marginally stable for the range of wavelengths for which the drift kinetic equation, which is used to derive Eq. (3.2), is valid, i.e., $k_\perp v_T \ll \omega_c$. From the above result, however, we can see that any effect that reduces the real frequency so that $\omega < \omega_e^*$ could lead to an instability, because Re $\sigma_e < 0$ for this range of frequency. One situation that can produce such an effect is a finite ion Larmor radius for a short wavelength, such that $k_\perp \sim \varrho_i^{-1} (= \omega_{ci}/v_{Ti})$. To take this effect into account, we have to go back to the Vlasov equation and recalculate the ion conductivity. If we do so, the ion conductivity changes to

$$\sigma_i = \frac{-i\omega\varepsilon_0\omega_{pi}^2}{k^2 v_{Ti}^2}\left[1+I_0(\lambda_i)\mathrm{e}^{-\lambda_i}\,\frac{\omega+\omega_i^*}{\sqrt{2}\,k_{\|}v_{Ti}}\,Z\left(\frac{\omega}{\sqrt{2}\,k_{\|}v_{Ti}}\right)\right], \qquad (3.3\,\mathrm{a})$$

which can be reduced for $\omega/k_{\|} \gg v_{Ti}$ to

$$\sigma_i = \frac{-i\omega\varepsilon_0\omega_{pi}^2}{k^2 v_{Ti}^2}\left(1 - \frac{\omega+\omega_i^*}{\omega} I_0(\lambda_i)\,\mathrm{e}^{-\lambda_i}\right) \tag{3.3b}$$

where

$$\lambda_i = \left(\frac{k_\perp v_{Ti}}{\omega_{ci}}\right)^2,$$

and I_0 is the modified Bessel function. If we solve the dispersion relation using the above ion conductivity,

$$\omega = \frac{\omega_i^*\,\mathrm{e}^{-\lambda_i} I_0(\lambda_i)}{(1+T_i/T_e) - \mathrm{e}^{-\lambda_i} I_0(\lambda_i)} \tag{3.4}$$

and the instability condition is obtained from Eqs. (3.4) and (2.139a) by setting Re $\sigma < 0$:

$$\left(1 + \frac{\omega_i^*}{\omega_e^*}\right)\mathrm{e}^{-\lambda_i} I_0(\lambda_i) - \left(1 + \frac{T_i}{T_e}\right) < 0. \tag{3.5}$$

Since electrons and ions have the same density gradient κ, $\omega_i^*/\omega_e^* = T_i/T_e$. Also, $\mathrm{e}^{-\lambda_i} I_0(\lambda_i) < 1$ for any finite value of the ion cyclotron radius $v_{Ti}/\omega_{ci}(\lambda_i \neq 0)$. Hence, the instability condition is always satisfied. This means that whenever a plasma has a density gradient it becomes unstable for a wave whose parallel phase velocity is between the thermal velocities of ions and electrons. For this reason this instability is called the universal instability.

Another effect that can produce an unstable solution is a temperature gradient which is opposite to the density gradient. As we can see from Eq. (2.138), the term that led to the drift wave frequency ω^* in Eq. (2.139b) originated from the term $\partial f_0/\partial x$. In the presence of a temperature gradient, this term produces, after integration in velocity space, a modification to ω_e^* such that $\omega_e^* \to \omega_e^*(1 - \frac{1}{2}\partial \ln T/\partial \ln n_0)$, because of the term v_{Te}^{-1} in front of the Z function. The resultant dispersion relation is then

$$1 - \frac{\omega_e^*}{\omega} + i\sqrt{\frac{\pi}{2}}\,\frac{\omega}{k_{\|} v_{Te}}\left[1 - \frac{\omega_e^*}{\omega}\left(1 - \frac{1}{2}\,\frac{\partial \ln T}{\partial \ln n_0}\right)\right] = 0. \tag{3.6}$$

Consequently an instability occurs when $\partial \ln T/\partial \ln n_0 < 0$, or when the temperature gradient and the density gradient have opposite signs (Rudakov and Sagdeev, 1961).

3.2b Medium β Case

Since plasmas in space usually have a fairly large value of β, the result obtained in the previous subsection is not applicable. When $\beta \sim (m_e/m_i)^{1/2}$, the plasma motion along the field line bends the magnetic line of force and excites shear Alfvén waves. Because the parallel phase velocity $\omega/k_{\|}$ of a shear Alfvén wave is given by $v_A = c\,\omega_{ci}/\omega_{pi} = v_{Te}(2m_e/m_i\beta_e)^{1/2}$ when $\beta \sim (m_e/m_i)^{1/2}$, the Alfvén speed becomes significantly smaller than the electron thermal speed and $k_{\|} v_A$ becomes comparable to the drift wave frequency.

If we take into account the coupling of the shear Alfvén wave with the drift wave, we can no longer assume an electrostatic perturbation. However, when β is still much smaller than unity, there is a way to simplify the involvement of the entire Maxwell's equations. Since this is a fairly useful technique for treating problems involving a medium β plasma we will look into it rather carefully. The idea is to decouple the compressional Alfvén mode by assuming $(\boldsymbol{\nabla} \times \boldsymbol{E})_{\|} = -\partial \boldsymbol{B}_{\|}/\partial t = 0$, that is taking into account only the effect of the field line bending. This assumption enables us to use a static potential for the perpendicular components of the wave electric field $\boldsymbol{E}_{\perp}$:

$$\boldsymbol{E}_{\perp} = -\boldsymbol{\nabla}_{\perp}\varphi. \tag{3.7}$$

For the parallel electric field we have to include the electromagnetic correction. Then Maxwell's equations give

$$k_{\perp}^2 \varphi + i k_{\|} E_{\|} = \varrho/\varepsilon_0, \tag{3.8a}$$

$$(i k_{\|}\varphi + E_{\|}) k_{\perp}^2 = i\omega \mu_0 J_{\|}. \tag{3.9}$$

If the quasi-neutrality condition is acceptable, we have instead of Eq. (3.8a),

$$\varrho \equiv \varrho_i + \varrho_e = 0. \tag{3.8b}$$

We now derive the charge density and the parallel current density for electrons, ϱ_e and $J_{\|e}$. For this we can again use the drift kinetic equation (2.129). Noting the contribution of the wave magnetic field, the linearized drift kinetic equation for electrons becomes

$$\frac{\partial f_1^{(e)}}{\partial t} + v_{\|}\frac{\partial f_1^{(e)}}{\partial z} + \left[-\frac{i k_{\perp}\varphi}{B_0}\left(1 - \frac{k_{\|} v_{\|}}{\omega}\right) + \frac{k_{\perp} v_{\|}}{\omega}\frac{E_{\|}}{B_0}\right]\frac{\partial f_0^{(e)}}{\partial x} - \frac{e}{m_e} E_{\|}\frac{\partial f_0^{(e)}}{\partial v_{\|}} = 0. \tag{3.10}$$

Solving for $f_1^{(e)}$, we have

$$f_1^{(e)} = \frac{k_\perp \varphi \kappa}{\omega B_0} f_0^{(e)} + \frac{\frac{k_\perp v_{||}}{\omega B_0} \kappa f_0^{(e)} + \frac{e}{m_e} \frac{\partial f_0^{(e)}}{\partial v_{||}}}{i(k_{||} v_{||} - \omega)} E_{||} \tag{3.11}$$

where κ is defined as in Eq. (2.140)

$$\kappa = -\frac{\partial \ln f_0}{\partial x}. \tag{2.140}$$

For the ions, we cannot assume an adiabatic relation and hence we have to use the Vlasov equation to obtain the perturbed distribution function $f_1^{(i)}$. We have then:

$$\begin{aligned} f_1^{(i)} = & \frac{e}{im_i} \sum_{\mathrm{n}} \sum_{\mathrm{m}} \frac{\mathrm{e}^{i(\mathrm{n}-\mathrm{m})\theta}}{\omega - k_{||} v_{||} - \mathrm{m}\,\omega_{ci}} J_{\mathrm{n}}\left(\frac{k_\perp v_\perp}{\omega_{ci}}\right) J_{\mathrm{m}}\left(\frac{k_\perp v_\perp}{\omega_{ci}}\right) \\ & \cdot \left[-ik_\perp \varphi \left\{ \left(1 - \frac{k_{||} v_{||}}{\omega}\right) \left(\frac{\partial f_0^{(i)}}{\partial v_\perp} \frac{\mathrm{n}\,\omega_{ci}}{k_\perp v_\perp} - \frac{\kappa}{\omega_{ci}} f_0^{(i)} \right) \right.\right. \\ & \left. + \frac{k_{||}}{k_\perp} \frac{\mathrm{n}\,\omega_{ci}}{\omega} \frac{\partial f_0^{(i)}}{\partial v_{||}} \right\} + E_{||} \left\{ \frac{k_\perp v_{||}}{\omega} \left(\frac{\partial f_0^{(i)}}{\partial v_\perp} \frac{\mathrm{n}\,\omega_{ci}}{k_\perp v_\perp} - \frac{\kappa}{\omega_{ci}} f_0^{(i)} \right) \right. \\ & \left.\left. + \left(1 - \frac{\mathrm{n}\,\omega_{ci}}{\omega}\right) \frac{\partial f_0^{(i)}}{\partial v_{||}} \right\} \right]. \end{aligned} \tag{3.12}$$

Note here that $f_0^{(i)}$ is a function of $x + v_y/\omega_{ci}$, $v_\perp$, and $v_{||}$. If we calculate the current and the charge density by integrating these perturbed distribution functions over velocity space and substitute the results into the wave equations (3.8b) and (3.9), we obtain the dispersion relation in the following form:

$$\begin{vmatrix} D_{\perp\perp} & D_{\perp ||} \\ D_{||\perp} & D_{||\,||} \end{vmatrix} = 0 \tag{3.13}$$

where $D_{ij} = D_{ij}^{(i)} + D_{ij}^{(e)}$ and

$$D_{\perp\perp}^{(e)} = \frac{-\kappa\,\omega_{pe}^2}{k_\perp \omega\,\omega_{ce}}$$

$$D_{\perp ||}^{(e)} = -\frac{\omega_{pe}^2}{k_\perp k_{||} v_{Te}^2} \left(1 - \frac{\omega_e^*}{\omega}\right) \left[1 + \int \frac{\omega f_0^{(e)}\, dv_{||}}{k_{||} v_{||} - \omega}\right]$$

$$D_{\|\perp}^{(e)} = -\frac{k_\perp k_\| c^2}{\omega^2}$$

$$D_{\|\|}^{(e)} = -\frac{\omega_{pe}^2}{k_\|^2 v_{Te}^2}\left(1-\frac{\omega_e^*}{\omega}\right)\left[1+\int\frac{\omega f_0^{(e)}\,dv_\|}{k_\| v_\| - \omega}\right]+\frac{c^2 k_\perp^2}{\omega^2}$$

$$D_{\perp\perp}^{(i)} = \frac{\omega_{pi}^2}{\omega k_\perp}\sum_{\mathrm{n}=-\infty}^{\infty}\int\frac{J_\mathrm{n}^2(k_\perp v_\perp/\omega_{ci})}{\mathrm{n}-a}\left\{a\left(\frac{\mathrm{n}\,\omega_{ci}}{k_\perp v_\perp}\frac{\partial f_0^{(i)}}{\partial v_\perp}-\frac{\kappa}{\omega_{ci}}f_0^{(i)}\right)\right.$$

$$\left.+\frac{\mathrm{n}\,k_\|}{k_\perp}\frac{\partial f_0^{(i)}}{\partial v_\|}\right\}2\pi v_\perp\,dv_\perp\,dv_\|$$

$$D_{\perp\|}^{(i)} = \frac{\omega_{pi}^2}{\omega_{ci} k_\perp}\sum_{\mathrm{n}=-\infty}^{\infty}\int\frac{J_\mathrm{n}^2(k_\perp v_\perp/\omega_{ci})}{\mathrm{n}-a}\left\{\frac{k_\perp v_\|}{\omega}\left(\frac{\mathrm{n}\,\omega_{ci}}{k_\perp v_\perp}\frac{\partial f_0^{(i)}}{\partial v_\perp}-\frac{\kappa}{\omega_{ci}}f_0^{(i)}\right)\right.$$

$$\left.+\left(1-\frac{\mathrm{n}\,\omega_{ci}}{\omega}\right)\frac{\partial f_0^{(i)}}{\partial v_\|}\right\}2\pi v_\perp\,dv_\perp\,dv_\|$$

$$D_{\|\perp}^{(i)} = \frac{\omega_{pi}^2}{\omega^2}\sum_{\mathrm{n}=-\infty}^{\infty}\int\frac{v_\| J_\mathrm{n}^2(k_\perp v_\perp/\omega_{ci})}{\mathrm{n}-a}\left\{a\left(\frac{\mathrm{n}\,\omega_{ci}}{k_\perp v_\perp}\frac{\partial f_0^{(i)}}{\partial v_\perp}-\frac{\kappa}{\omega_{ci}}f_0^{(i)}\right)\right.$$

$$\left.+\frac{\mathrm{n}\,k_\|}{k_\perp}\frac{\partial f_0^{(i)}}{\partial v_\perp}\right\}2\pi v_\perp\,dv_\perp\,dv_\|$$

$$D_{\|\|}^{(i)} = \frac{\omega_{pi}^2}{\omega\omega_{ci}}\sum_{\mathrm{n}=-\infty}^{\infty}\int\frac{v_\| J_\mathrm{n}^2(k_\perp v_\perp/\omega_{ci})}{\mathrm{n}-a}\left\{\frac{k_\perp v_\|}{\omega}\left(\frac{\mathrm{n}\,\omega_{ci}}{k_\perp v_\perp}\frac{\partial f_0^{(i)}}{\partial v_\perp}-\frac{\kappa}{\omega_{ci}}f_0^{(i)}\right)\right.$$

$$\left.+\left(1-\frac{\mathrm{n}\,\omega_{ci}}{\omega}\right)\frac{\partial f_0^{(i)}}{\partial v_\|}\right\}2\pi v_\perp\,dv_\perp\,dv_\|$$

where

$$a = \frac{\omega - k_\| v_\|}{\omega_{ci}}.$$

The dispersion relation obtained above has a complicated structure. When solved for $\omega \gtrsim \omega_{ci}$ it produces a finite β modification to the drift cone instability discussed in Subsection 2.3c.

One can also derive the *drift cyclotron instability* when $\omega_e^* > \mathrm{n}\,\omega_{ci}$. This instability occurs as a result of the coupling between the electron drift wave and the ion Bernstein wave. The instability is essentially electrostatic, hence might have been included in the previous subsection.

The dispersion relation for $\omega \sim n\omega_{ci}$ is obtained by letting $c \to \infty$ in Eq. (3.13),

$$2 + \frac{\omega + \omega_i^*}{\sqrt{2}\, k_{\|} v_{Ti}} Z\left(\frac{\omega - n\omega_{ci}}{\sqrt{2}\, k_{\|} v_{Ti}}\right) e^{-\lambda_i} I_0(\lambda_i) + \frac{\omega - \omega_e^*}{\sqrt{2}\, k_{\|} v_{Te}} Z\left(\frac{\omega}{\sqrt{2}\, k_{\|} v_{Te}}\right) = 0, \tag{3.14}$$

where

$$\lambda_i = k_\perp^2 v_{Ti}^2 / \omega_{ci}^2 .$$

The instability occurs even for a flute mode ($k_{\|} = 0$) and the condition of the instability is given by [Mikhailovskii and Timofeev (1963)],

$$k\varrho_i \geq 2n\left(\frac{m_e}{m_i} + \frac{v_A^2}{c^2}\right)^{1/2}, \tag{3.15}$$

where ϱ_i is the ion Larmor radius and n is the nth harmonic of the ion cyclotron frequency.

Let us now look at the effect of finite β on a drift wave instability of the type discussed in the previous subsection. For $\omega \ll \omega_{ci}$ and $v_{Ti} \ll \omega / k_{\|} \ll v_{Te}$, the components of D_{ij} in Eq. (3.13) reduce to

$$\left.\begin{aligned}
D_{\perp\perp} &= -\frac{\omega_{pi}^2}{k_\perp^2 v_{Ti}^2}\left[1 - e^{-\lambda_i} I_0(\lambda_i)\right]\left(1 + \frac{\omega_i^*}{\omega}\right) \\
D_{\perp\|} &= \frac{\omega_{pi}^2}{\omega^2}\frac{k_{\|}}{k_\perp} e^{-\lambda_i} I_0(\lambda_i)\left(1 + \frac{\omega_i^*}{\omega}\right) \\
&\quad - \frac{\omega_{pe}^2}{k_\perp k_{\|} v_{Te}^2}\left(1 - \frac{\omega_e^*}{\omega}\right)\left(1 + i\sqrt{\frac{\pi}{2}}\frac{\omega}{k_{\|} v_{Te}}\right) \\
D_{\|\perp} &= -\frac{k_\perp k_{\|} c^2}{\omega^2} \\
D_{\|\|} &= \frac{\omega_{pi}^2}{\omega^2} e^{-\lambda_i} I_0(\lambda_i)\left(1 + \frac{\omega_i^*}{\omega}\right) \\
&\quad - \frac{\omega_{pe}^2}{k_{\|}^2 v_{Te}^2}\left(1 - \frac{\omega_e^*}{\omega}\right)\left(1 + i\sqrt{\frac{\pi}{2}}\frac{\omega}{k_{\|} v_{Te}}\right) + \frac{c^2 k_\perp^2}{\omega^2},
\end{aligned}\right\} \tag{3.16}$$

where we have assumed isotropic Maxwell distributions for both species. If we substitute these expressions into Eq. (3.13), we can obtain the drift wave dispersion relation for a finite β plasma. Assuming $T_e \gtrsim T_i$, the ion contributions to $D_{\perp\|}$ and $D_{\|\|}$ become negligible in view of $\omega \gg k_\| v_{Ti}$ and the resultant dispersion relation simplifies to

$$(\omega^2+\omega\omega_i^*-k_\|^2 v_A^2)\left(1-\frac{\omega_e^*}{\omega}\right)\left(1+i\sqrt{\frac{\pi}{2}}\,\frac{\omega}{k_\| v_{Te}}\right)=\frac{T_e}{T_i}\,\lambda_i k_\|^2 v_A^2\left(1+\frac{\omega_i^*}{\omega}\right). \tag{3.17a}$$

We note that if we let $c \to \infty$, (thus $v_A \to \infty$) in this expression, we can recover the dispersion relation of the electrostatic drift wave. We also note that if we take the limit of zero Larmor radius by letting $\lambda_i \to 0$, we have three real roots,

$$\omega_1=\omega_e^*, \tag{3.18}$$

$$\omega_2=\frac{-\omega_i^*+\sqrt{(\omega_i^*)^2+4\omega_A^2}}{2}, \tag{3.19}$$

and

$$\omega_3=\frac{-\omega_i^*-\sqrt{(\omega_i^*)^2+4\omega_A^2}}{2} \tag{3.20}$$

where $\omega_A = k_\| v_A$. Therefore the finite β effect produces two additional modes, one of which corresponds to the ion drift mode (ω_3) and the other to the shear Alfvén mode (ω_2). The nature of the instability is modified correspondingly. For example, if β is small such that $2\omega_A > \omega_i^*$, the instability occurs at

$$\omega \sim \omega_e^* \tag{3.21}$$

and the growth rate γ is approximately given by

$$\gamma=\lambda_i\omega_e^*\sqrt{\frac{\pi}{2}}\,\frac{\omega_e^*}{k_\| v_{Te}}\left(1+\frac{T_e}{T_i}\right). \tag{3.22}$$

However, if β is increased such that $2\omega_A < \omega_i^*$, the shear Alfvén wave becomes unstable, with the frequency of the instability given by

$$\omega \sim \frac{\omega_A^2}{\omega_i^*} \tag{3.23}$$

and the corresponding growth rate by

$$\gamma=\lambda_i \frac{\omega_A^2}{\omega_i^*} \sqrt{\frac{\pi}{2}} \frac{\omega_A^2}{k_{\|} v_{Te} \omega_i^*}. \tag{3.24}$$

The coupling to the Alfvén mode is a consequence of the bending of the magnetic field line due to the motion of electrons along the field line.

An important consequence of the finite β effect is the reduction of the frequency of the instability (as seen from Eq. (3.23)) and the resultant increase in the ion Landau damping due to the lowered phase velocity of the wave. In fact it was shown by Mikhailovskaya and Mikhailovskii (1963) that when $\beta \gtrsim 0.13$, the drift wave instability is completely stabilized by the ion Landau damping.

3.2c High β Case

When the plasma pressure increases further and β becomes comparable to unity, the compressional Alfvén wave becomes coupled with the drift wave and the instability is generated either as a result of the inversed transit time damping or the resonant interaction of ions in their gradient magnetic field drift.

This fact can be seen because $\beta_i=2 v_{Ti}^2/v_A^2$, hence $k_\perp v_A=k_\perp v_{Ti} \sqrt{2/\beta_i}= =\omega_i^* (\omega_{ci}/\kappa v_{Ti}) \sqrt{2/\beta_i}$ thus $k_\perp v_A$ becomes comparable to ω_i^* when $\beta_i \sim 0(1)$ and $\kappa \varrho_i \sim 0(1)$.

Another important effect of a large β plasma arises from the fact that the particle drifts due to the gradient in magnetic flux density or the curvature of the field line becomes comparable to the diamagnetic drift. For example, if we consider a case of a straight magnetic field, and plasma nonuniformity in the y direction, the pressure balance conditions give

$$\frac{d}{dy}\left(\frac{B^2}{2\mu_0}+\sum_{i,e} n_0 T_j\right)=0. \tag{3.25}$$

If we introduce a new drift wave frequency due to the gradient of the magnetic field,

$$\omega_j^G=\frac{k_\perp v_{Tj}^2}{\omega_{cj}} \frac{d \ln B}{dy}, \tag{3.26}$$

from Eq. (3.25) ω^G is related to ω^* by

$$2\omega_j^G+\beta\left(1+\frac{\partial \ln T}{\partial \ln n_0}\right)\omega_j^*=0 \tag{3.27}$$

where $j=i, e$ and β and T are the total $\beta(=\beta_i+\beta_e)$ and temperature (T_i+T_e), respectively (ω^* is defined here by using $\partial \ln n_0/\partial y$). Thus if $\beta \sim 0(1)$, ω^G becomes comparable to ω^* and has an opposite sign.

Let us consider drift wave instabilities under these circumstances. As an example, we will consider a situation inside the magnetosphere, where there exists a certain amount of cold plasma concentration. As shown in Subsection 2.4b, the dispersion relation of a compressional mode can then be written as

$$\frac{c^2 k^2}{\omega^2} - \varepsilon_{yy} = 0. \tag{2.162}$$

This expression is derived for a wave propagating in the (x, z) plane with a magnetic field in the z direction. Consequently we take the direction of nonuniformity to be the y direction: $\boldsymbol{B}_0 = B(y)\boldsymbol{e}_z$, $n_0 = n_0(y)$, so that the k vector has a component in the direction of the drift. ε_{yy} can be obtained as before by integrating the Vlasov equation along the unperturbed trajectory. In this case, the unperturbed trajectory must include the drift associated with $\nabla B_0 = dB_0/dy\, \boldsymbol{e}_y$. ε_{yy} then becomes (Hasegawa, 1971)

$$\varepsilon_{yy} = -\sum_{i,e} \frac{\omega_{pj}^2}{\omega} \sum_{\mathrm{n}=0,1} \int \frac{v_\perp^2 [J_{\mathrm{n}}(k_\perp v_\perp/\omega_{cj})]^2}{\omega - \mathrm{n}\omega_{cj} - k_{||} v_{||} + k_\perp v_{Bj}} \left(\frac{m_j}{T_j} + \frac{k_\perp}{\omega \omega_{cj}} \frac{d}{dy} \right) f_0\, d\boldsymbol{v} \tag{3.28}$$

where v_B is the drift speed due to the gradient of the magnetic flux density, given by

$$v_{Bj} = \frac{v_\perp^2}{2\omega_{cj}} \frac{d \ln B_0}{dy}. \tag{3.29}$$

When we expand the Bessel function in the above expression for a long wavelength perturbation and assume that the cold plasma density significantly exceeds that of the hot plasma, we have the following dispersion relation,

$$1 - \frac{\omega^2}{k^2 v_A^2} + \sum_{i,e} \frac{\beta_j k_\perp^2}{2k^2} \left\{ \left[\omega + \omega_j^* \left(1 - \frac{3}{2}\eta_j \right) \right] I_{1j} + \omega_j^* \eta_j I_{2j} \right\} = 0 \tag{3.30}$$

v_A is the Alfvén speed in the cold plasma which is assumed to be uniform and

$$\eta_j = \frac{\partial \ln T_j}{\partial \ln n_0}, \tag{3.31}$$

$$I_{1j}=\frac{1}{\sqrt{\pi}}\int_0^\infty dw \int_{-\infty}^{\infty} du \frac{w^2 \exp(-w-u^2)}{\omega-\sqrt{2}\,k_{\|}v_{Tj}u+\omega_j^G w}, \tag{3.32}$$

$$I_{2j}=\frac{1}{\sqrt{\pi}}\int_0^\infty dw \int_{-\infty}^{\infty} du \frac{w^2 (w+u^2) \exp(-w-u^2)}{\omega-\sqrt{2}\,k_{\|}v_{Tj}u+\omega_j^G w}, \tag{3.33}$$

$$\omega_j^*=\frac{k_\perp v_{Tj}^2}{\omega_{cj}}\frac{d\ln n_0}{dy} \tag{2.140}$$

and

$$\omega_j^G=\frac{k_\perp v_{Tj}^2}{\omega_{cj}}\frac{d\ln B_0}{dy}. \tag{3.26}$$

n_0 in these expressions indicates the hot plasma density.

As we can see from the integrals in Eqs. (3.32) and (3.33), the wave particle resonant interaction occurs at $\omega=k_{\|}v_{\|}$ and $\omega=-\omega^G$. The former resonance corresponds to the transit time damping which is discussed in Subsection 2.4b. The resonance at $\omega=-\omega_G$ is a new resonance which originates from particles whose speed across the magnetic field due to gradient B_0 drift coincides with the wave phase velocity in the same direction. When $k_{\|}\neq 0$, the instability occurs due to the particles with $v_{\|}=\omega/k_{\|}$; for a flute mode ($k_{\|}=0$), the instability is caused by the particles whose gradient B_0 drift speed is in resonance with the wave phase velocity, $\omega^G\sim-\omega$. In either case, the condition of instability is given by (Hasegawa, 1971),

$$\frac{n_0}{n_{\text{cold}}}<\beta\kappa_0^2\varrho_i^2/2 \tag{3.34}$$

where κ_0 is the measure of density, temperature or magnetic field gradients, ϱ_i is the ion Larmor radius of the hot component and n_{cold} is the cold plasma density. Consequently we see that, even if the ordinary drift wave instability is stabilized for a high β plasma, a compressional Alfvén wave can be excited in a plasma with $\beta\gtrsim n_0/n_{\text{cold}}$.

3.2d Notes on the Application of Drift Wave Instabilities

The theories of drift wave instabilities presented in the foregoing subsections may give the impression that a nonuniform plasma tends to be unstable under almost any circumstances. However, when we apply the results to an actual plasma, we find many factors that produce stabilization effects.

Instead of giving specific examples of possible occurrences of a drift wave instability in space, we will point out here effects which should be considered when applying the drift wave instability to space plasmas.

First, we should note the effect of finite β stabilization due to ion Landau damping for $\beta \gtrsim 0.13$. Except for plasma in the plasmapause or in the ionosphere, most space plasma has a value of β comparable or larger than 0.1. Hence this is an important point to check. However, if $\beta \geq 0(1)$ or $\beta \gtrsim n_{\text{hot}}/n_{\text{cold}}$ one can consider a drift wave excitation of a compressional Alfvén wave.

The second effect to consider arises from an admixture of cold electrons. As we can see from the derivation of the drift wave instability in Subsection 3.2a, the inversed hot electron Landau damping of a parallel electrostatic potential drives the instability. However, if even a small amount of cold electrons exists, they tend to short circuit such a parallel potential very quickly due to their small inertia.

If cold electrons with density n_c are mixed in, the dispersion relation (3.17a) is modified to

$$(\omega^2 + \omega\omega_i^* - k_{\|}^2 v_A^2)\left[\left(1 - \frac{\omega_e^*}{\omega}\right)\left(1 + i\sqrt{\frac{\pi}{2}}\,\frac{\omega}{k_{\|} v_{Te}}\right) - \frac{n_c}{n_0}\,\frac{k_{\|}^2 v_{Te}^2}{\omega^2}\right] \tag{3.17b}$$

$$= \frac{T_e}{T_i}\,\lambda_i k_{\|}^2 v_A^2 \left(1 + \frac{\omega_i^*}{\omega}\right).$$

The modification appears as the last term in the left hand side. Since we have to assume $\omega \ll k_{\|} v_{Te}$ for the frequency range of the drift wave instability, the additional term, $n_c k_{\|}^2 v_{Te}^2 / n_0 \omega^2$ produces a large modification to the dispersion relation even if $n_c < n_0$. From this expression it can be shown that a fractional admixture $(\sim \sqrt{m_e/m_i})$ of cold electrons stabilizes the electrostatic drift wave instability for the case $k_{\|} v_A > \omega_i^*$, while for the case $k_{\|} v_A < \omega_i^*$, it reduces the growth rate of the drift wave instability of a shear Alfvén wave to an insignificant level.

The third effect we should note is the effect of the bounce motion of electrons in a magnetic mirror. When we consider long wavelength perturbations in the direction parallel to the magnetic field which are comparable to the length of a field line, such as perturbations inside the magnetosphere, $k_{\|} v_{Te}$ becomes the bounce frequency of electrons ω_{be}. As shown in Subsection 2.2c, if $\omega < \omega_{be}$ (hence $\omega < k_{\|} v_{Te}$), the Landau damping disappears because of the periodic motion of electrons. When this happens the driving term of the drift wave instability $i\sqrt{\pi}\,\omega/(k_{\|} v_{Te}\sqrt{2})$ disappears and so does the instability. In this circumstance, some other dissipation mechanism is required to excite the drift wave instability.

Hagege *et al.* (1973) have proposed that whistler turbulence may produce this dissipation and derived the drift wave instability for $\omega < \omega_{be}$.

Although many theoretical attempts have been made to apply a drift wave instability to magnetospheric plasmas, most of them lack a careful check as to the applicability of the theory. As has been shown here, the drift wave instability is very sensitive to plasma parameters and it is desirable to take all the realistic factors into consideration when applying it to magnetospheric plasmas.

3.3 Rayleigh-Taylor Instability

This is a classical instability of a fluid resting on a lighter fluid. Because the instability is driven by the gravitational force, it is also called the gravitational instability. For a plasma on a curved field line, the centrifugal force on particles moving along the field line acts like a gravitational force. This force may produce an instability at the outer boundary of the plasma where the density decreases radially. This instability is often called the flute instability because the most unstable mode has a flute-like perturbation, which propagates perpendicular to the magnetic field (a field aligned perturbation, Rosenbluth and Longmire, 1957).

Instead of considering a curved field line, we simulate the centrifugal force by gravitational force, a force which is independent of the sign of the charge of a particle and proportional to the mass, and assume a planar geometry. We take the gravitational field $\boldsymbol{g}$ to be in the positive x direction, the direction of decreasing density, i.e.,

$$\boldsymbol{g} \cdot \nabla n_0 < 0 . \qquad (3.35)$$

We assume for simplicity that the plasma is cold and collisionless, and consider an electrostatic perturbation that propagates in the y-z plane, where z is the direction of the magnetic field. Normally, when a gravitational instability is discussed, one considers perturbation only in the y direction, normal to both the magnetic field and the density gradient. Here we will consider propagation also in the direction of the magnetic field to show an important stabilization effect.

The linearized equation of motion for ions takes the form

$$-i\omega' \boldsymbol{v}_{\perp 1} = \frac{e}{m_i}(-i\boldsymbol{k}_\perp \varphi + \boldsymbol{v}_{\perp 1} \times \boldsymbol{B}_0) \qquad (3.36)$$

where

$$\omega' = \omega + k_\perp v_g = \omega + \frac{k_\perp g}{\omega_{ci}} \qquad (3.37)$$

and $k_\perp = k_y$, $k_{||} = k_z$.

Solving Eq. (3.36) for $\omega \ll \omega_{ci}$

$$\boldsymbol{v}_{\perp 1} = i\,\frac{\boldsymbol{B}_0 \times \boldsymbol{k}_\perp}{B_0^2}\,\varphi - \frac{\omega'}{B_0\,\omega_{ci}}\,\boldsymbol{k}_\perp \varphi\,. \tag{3.38}$$

Substituting this expression into the linearized equation of continuity, we have

$$\left(\frac{n_1}{n_0}\right)_{\text{ion}} = \varphi\left[\frac{\kappa k_\perp}{B_0\,\omega'} + \frac{e}{m_i}\left(\frac{k_\parallel^2}{\omega^2} - \frac{k_\perp^2}{\omega_{ci}^2}\right)\right] \tag{3.39}$$

where

$$\kappa = -\frac{d \ln n_0}{dx} > 0\,.$$

For electrons, because the gravitational force produces little drift (smaller by the mass ratio than that of the ions; note however that the electron drift can become comparable to the drift of the ions if centrifugal force is used), we have

$$\left(\frac{n_1}{n_0}\right)_{\text{electrons}} = \varphi\left(\frac{\kappa k_\perp}{B_0\,\omega} - \frac{e}{m_e}\,\frac{k_\parallel^2}{\omega^2}\right). \tag{3.40}$$

The dispersion relation is obtained by assuming quasi-neutrality to be

$$\frac{k\,\omega_{ci}}{k_\perp}\left(\frac{1}{\omega + k_\perp g/\omega_{ci}} - \frac{1}{\omega}\right) = 1 - \frac{m_i}{m_e}\left(\frac{k_\parallel\,\omega_{ci}}{k_\perp\,\omega}\right)^2. \tag{3.41}$$

Now if we assume an ideal flute mode such that $k_\parallel = 0$, the unstable solution is easily obtained by assuming $\omega \gg k_\perp g/\omega_{ci}$ and expanding the first term of the left-hand side of expression (3.41) in powers of $k_\perp g/(\omega_{ci}\,\omega)$ as

$$\omega^2 + g\kappa = 0 \tag{3.42a}$$

or

$$\omega = \pm i (g\kappa)^{1/2}. \tag{3.42b}$$

Eq. (3.42) corresponds to the growth rate of the classical gravitational instability.

However, unlike classical fluids, the anisotropic nature of the plasma conductivity produces an interesting stabilization effect if $k_\parallel \neq 0$. For example, as in the case of magnetospheric plasma, if the field lines are connected to a conductive region perpendicular to themselves (the ionosphere), the charge separation perpendicular to the field lines that drives the instability is short-circuited by the large electron conductivity in the parallel direction in a way similar to the case of the drift-wave instability. This effect is represented by the second term on the right-hand side of expression (3.41). One can see from the previous argument that the unstable solution disappears when the right-hand side changes its sign. The stability condition then is obtained roughly as *

$$\frac{k_{\|}}{k_{\perp}} > \left(\frac{m_e}{m_i}\right)^{1/2} \frac{(g\kappa)^{1/2}}{\omega_{ci}}. \tag{3.43}$$

Eq. (3.43) shows that only a perturbation having small perpendicular wavelength (large $k_{\perp}$) grows under such circumstances.

On the other hand, it is known that the gravitational instability is stabilized for a perpendicular wavelength which is short compared to the ion cyclotron radius (Lehnert, 1961) because of the neutralization of charge separation due to the finite size of the ion cyclotron radius. The stabilization condition due to this finite cyclotron-radius effect can be obtained by using the Vlasov equation including the gravitational force. The stabilization condition then reads

$$g\kappa < (\omega_i^*)^2/4 \tag{3.44a}$$

where ω_i^* is the ion drift-wave frequency. Using $g = T_i/m_i R$ where R is the radius of the curvature of the field line, and T_i is the ion temperature in energy units, Eq. (3.44a) can be expressed also as

$$k_{\perp}^2 \varrho_i^2 > 4/(\kappa R) \tag{3.44b}$$

where $\varrho_i (= v_{Ti}/\omega_{ci})$ is the ion Larmor radius. Therefore, the gravitational instability is stabilized for a long wavelength perturbation by the short-circuiting of electrons moving rapidly parallel to the field lines, and for a short wavelength perturbation by the finite cyclotron radius effect of ions.

If one assumes a model in which the ionosphere is a perfect conductor in a low frequency regime ($\omega \ll \omega_{ci}$), one can see that by combining Eqs. (3.43) and (3.44b) gravitational instability is stabilized for a perturbation of *any size* if

$$R k_{\|} > 2 \left(\frac{m_e}{m_i}\right)^{1/2}. \tag{3.45}$$

* Because we used the cold electron model, this expression is valid only for $\omega > k_{\|} v_{Te}$. For the magnetosphere, $g = T_i/m_i R$, thus $\omega/k_{\|} \sim v_{Ti}$, hence this result is applicable when the hot ion thermal speed, v_{Ti}, is larger than the cold electron thermal speed, v_{Te}. Otherwise ω^2 in Eq. (3.40) should be replaced by $(-k_{\|}^2 v_{Te}^2)$ and the electrons contribute to *reduce* the growth rate instead of complete stabilization.

The parallel wave number $k_{||}$ is lower-bounded by π over twice the length of the field line ($\sim \pi/\mathrm{L}\, R_E$ where L is the equatorial crossing distance in units of earth radii, R_E). Thus $k_{||} > \pi/\mathrm{L}\, R_E$, whereas the radius of curvature $R \sim \mathrm{L}\, R_E/3$. Hence Eq. (3.45) is always satisfied. This result indicates that the magnetospheric plasma may be stable against the gravitational instability. Chang *et al.* (1965, 1966) have taken into account effects of the realistic ionospheric conductivity as well as the curvature of the field line to study the Rayleigh-Taylor instability in the ring current plasma. Using kinetic equations to introduce the effect of a parallel electric field, they have concluded that if adiabatic interchange is assumed, the critical radial dependence of the energy becomes r^{-7}. However, if the condition of adiabaticity is violated by a wave-particle resonance, the sufficient condition for stability is shown to be, considering the ionospheric conductivity,

$$\left(\frac{\Delta r}{R_E}\right) \frac{\mathrm{m}}{\mathrm{L}^4} \frac{n_i}{n_0} \gtrsim 0.3 \tag{3.46}$$

where $\Delta r \sim 50$ km, R_E (=earth's radius) ~ 6.300 km, m is the azimuthal mode number (integer), L is the radial distance in units of earth radii, n_i is the ionospheric plasma density, and n_0 is the proton density of the ring current. They concluded that the stability condition can be easiliy satisfied for the day side because of larger n_i, but may be marginal for the night side.

3.4 Kelvin-Helmholtz Instability

Another classical fluid instability which is applicable to plasma dynamics is the Kelvin-Helmholtz instability, which is excited by a velocity shear. We will present two examples of this instability which are relevant in magnetospheric dynamics. One is the hydromagnetic instability which may be excited at the boundary of the magnetosphere by the solar wind flow. The other is the electrostatic Kelvin-Helmholtz instability excited by $\boldsymbol{E} \times \boldsymbol{B}$ flow in a non-uniform electric field in the auroral sheet.

3.4a Hydromagnetic Instability

We consider a hydromagnetic perturbation near a boundary of discontinuous flow velocity. We take planar geometry and the coordinate system as shown in Fig. 27, where y, z are along the boundary surface, and x is normal to the boundary. Because microscopically such a bound-

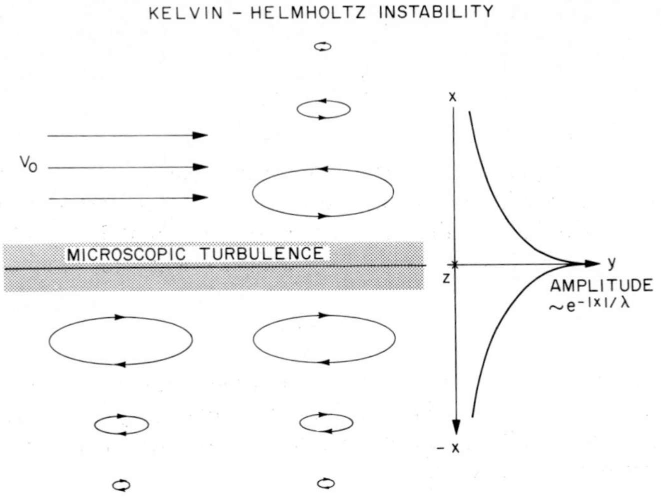

Fig. 27. Coordinate system employed for hydromagnetic Kelvin-Helmholtz instability. At $x>0$, there exists a flow of the plasma with $\boldsymbol{v}=\boldsymbol{v}_0$

ary may not be under dynamic equilibrium (Lerche, 1967; Parker, 1967 a, b, 1968 a, b), there may exist at the boundary some kind of turbulent layer with a thickness of the order of the proton Larmor radius. In this text, however, we derive the Kelvin-Helmholtz instability for the most ideal case to demonstrate its physical properties; thus, we assume a smooth boundary in MHD scale and also assume incompressible ideal MHD perturbations (Chandrasekhar, 1961; Sen, 1963).

If we introduce a new vector ξ which corresponds to the displacement of the plasma fluid from equilibrium, the linearized MHD equations (1.20), (1.25) and (1.27) can be expressed, after eliminating the electric field, as

$$m_i n_0 \ddot{\xi} = \frac{1}{\mu_0} \left[(\nabla \times \boldsymbol{B}_1) \times \boldsymbol{B}_0 + (\nabla \times \boldsymbol{B}_0) \times \boldsymbol{B}_1 \right] - \nabla p_1 , \tag{3.47}$$

$$\boldsymbol{B}_1 = \nabla \times (\xi \times \boldsymbol{B}_0) = (\boldsymbol{B}_0 \cdot \nabla)\xi - (\xi \cdot \nabla)\boldsymbol{B}_0 - \boldsymbol{B}_0 (\nabla \cdot \xi) \tag{3.48}$$

where

$$\frac{\partial \xi}{\partial t} = \boldsymbol{v}_1 . \tag{3.49}$$

After standard vector operations together with $\nabla \cdot \boldsymbol{B}_1 = \nabla \cdot \boldsymbol{B}_0 = 0$, we have from Eq. (3.47),

$$\nabla^2 \tilde{p}_1 = \nabla \cdot \left[(\boldsymbol{B}_0 \cdot \nabla)\boldsymbol{B}_1 + (\boldsymbol{B}_1 \cdot \nabla)\boldsymbol{B}_0 \right] / \mu_0 - \nabla \cdot (m_i n_0 \ddot{\xi}) \tag{3.50}$$

where $\tilde{p}_1$ is the total pressure given by

$$\tilde{p}_1 = \frac{\boldsymbol{B}_1 \cdot \boldsymbol{B}_0}{\mu_0} + p_1 . \tag{3.51}$$

If we assume that the unperturbed plasma density and the magnetic flux density are uniform (except at the boundary surface), and that the perturbation can be regarded as incompressible so that $\nabla \cdot \boldsymbol{v}_1 = \nabla \cdot \boldsymbol{\xi} = 0$, Eq. (3.50) immediately produces the necessary wave equation

$$\nabla^2 \tilde{p}_1 = 0 . \tag{3.52}$$

Because this is the Laplace equation, if we assume a wave-like perturbation along the boundary surface, i.e., $\exp i(k_y y + k_z z - \omega t)$, the solution exponentially decays away from the surface. Such a perturbation is called a surface wave:

$$\tilde{p}_1 = p_0 \, \mathrm{e}^{-\kappa|x|} \, e^{i(k_y y + k_z z - \omega t)} \tag{3.53}$$

where

$$\kappa = (k_y^2 + k_z^2)^{1/2} . \tag{3.54}$$

The plasma displacement $\boldsymbol{\xi}$ corresponding to this pressure perturbation is obtained by combining Eq. (3.48) with (3.47) and

$$[\mu_0 m_i n_0 \omega^2 - (\boldsymbol{k} \cdot \boldsymbol{B}_0)^2] \boldsymbol{\xi} = \mu_0 \nabla \tilde{p}_1 + \boldsymbol{c} \tag{3.55}$$

where

$$\boldsymbol{c} = (\boldsymbol{B}_0 \cdot \nabla) \, [(\boldsymbol{\xi} \cdot \nabla) \boldsymbol{B}_0 + \boldsymbol{B}_0 (\nabla \cdot \boldsymbol{\xi})] - (\boldsymbol{B}_1 \cdot \nabla) \boldsymbol{B}_0 . \tag{3.56}$$

The quantity $\boldsymbol{c}$, which represents the coupling between the surface wave and the shear Alfvén wave (represented zero on the left hand side of Eq. (3.55)), vanishes in a uniform plasma. In that case, $\boldsymbol{\xi}$ can be expressed immediately as

$$\boldsymbol{\xi} = \frac{\nabla \tilde{p}_1}{m_i n_0 [\omega^2 - (\boldsymbol{k} \cdot \boldsymbol{v}_A)^2]} \tag{3.57}$$

where $\boldsymbol{v}_A = \boldsymbol{B}_0 / (\mu_0 m_i n_0)^{1/2}$ and $\boldsymbol{k} = k_z \boldsymbol{e}_z + k_y \boldsymbol{e}_y$. The dispersion relation can be obtained by applying suitable boundary conditions to the solution obtained in Eqs. (3.53) and (3.57). We take these boundary conditions to be the continuity of the total pressure $\tilde{p}_1$ and of the normal component of $\boldsymbol{\xi}$ i.e., ξ_x. We represent quantities in $x > 0$ by subscript I and those in $x < 0$ by subscript II. We note then that ω_{I} is Doppler shifted with respect to ω_{II}, i.e., $\omega_{\mathrm{I}} = \omega - \boldsymbol{k} \cdot \boldsymbol{v}_0$, $\omega_{\mathrm{II}} = \omega$. The continuity of ξ_x from Eq. (3.57) gives, after using the fact that $\tilde{p}_1$ is continuous,

$$\frac{1}{n_{0\mathrm{I}}[(\omega-\boldsymbol{k}\cdot\boldsymbol{v}_0)^2-(\boldsymbol{k}\cdot\boldsymbol{v}_A)_{\mathrm{I}}^2]}+\frac{1}{n_{0\mathrm{II}}[\omega^2-(\boldsymbol{k}\cdot\boldsymbol{v}_A)_{\mathrm{II}}^2]}=0. \quad (3.58)$$

The condition of instability can then be obtained from the condition of a complex root of ω as

$$(\boldsymbol{k}\cdot\boldsymbol{v}_0)^2>\left(\frac{1}{n_{0\mathrm{I}}}+\frac{1}{n_{0\mathrm{II}}}\right)[n_{0\mathrm{I}}(\boldsymbol{k}\cdot\boldsymbol{v}_A)_{\mathrm{I}}^2+n_{0\mathrm{II}}(\boldsymbol{k}\cdot\boldsymbol{v}_A)_{\mathrm{II}}^2]. \quad (3.59)$$

This expression presents various properties of the Kelvin-Helmholtz instability of a MHD wave. First, and most obvious, is that the instability is the consequence of a relative drift $\boldsymbol{v}_0$ of the two fluids along the discontinuous boundary. Second, the instability occurs more easily for a $\boldsymbol{k}$ vector perpendicular to the unperturbed magnetic field $\boldsymbol{B}_0$ and parallel to the flow velocity $\boldsymbol{v}_0$. This means that the instability is more easily excited when the plasma flow is perpendicular to the magnetic field. Third, the perturbed fluid motion is in elliptically polarized eddies as shown in Fig. 27. This can be seen by taking the ratio of the x and y components of $\boldsymbol{\xi}$. If we assume $k_z \ll k_y$, $\nabla\cdot\boldsymbol{\xi}=0$ gives $-\kappa\,\xi_x+ik_y\xi_y\sim 0$ hence $\xi_x/\xi_y=ik_y/\kappa=ik_y/(k_y^2+k_z^2)^{1/2}$. Thus the $\boldsymbol{\xi}$ vector is elliptically polarized in the x-y plane with an ellipticity of $(k_y^2+k_z^2)^{1/2}/k_y$.

When the Kelvin-Helmholtz instability is applied to the magnetospheric boundary, the incompressibility assumption becomes invalid because the solar wind speed has a high Alfvén Mach number, and we must consider compressible perturbations (Sen, 1964; Southwood, 1968; Ong and Roderick, 1972). The instability condition becomes more complex. In this case, the magnetosonic wave is also excited. We should also take into account the effect of a turbulent layer which may appear as a consequence of the microscopically nonequilibrium nature of the boundary layer (Eviatar and Wolf, 1968); however, this effect has not been considered yet.

The Kelvin-Helmholtz instability at the magnetopause is considered to be one of the major causes of pc 3 to pc 5 magnetic pulsations. Commonly observed elliptically polarized pulsations with periods of 30 sec to a few minutes are considered to be the result of field line oscillations of a shear Alfvén wave in resonance with the frequency of the Kelvin-Helmholtz instability (Chen and Hasegawa, 1974). Boller and Stolov (1970) discussed the semiannual variation of geomagnetic activity in terms of the excitation of the Kelvin-Helmholtz instability. As shown by Eq. (3.59), the condition for the instability changes as a function of the angle between $\boldsymbol{v}_0$ and $\boldsymbol{B}_0$; the tilt of the dipole axis which changes semiannually consequently affects the occurrence of the instability and also any geomagnetic activity caused by the instability.

3.4b Electrostatic Instability

Here we consider an example of the electrostatic Kelvin-Helmholtz instability produced by shear flow due to an E_0 (nonuniform) cross B_0 (uniform) drift of a plasma. The nonuniform electric field is a consequence of a non-neutral, charged sheet as shown in Fig. 28. As an example, we consider an electron sheet with thickness $2a$ with its surface parallel to the uniform magnetic field. Because of the negative charge of electrons, a steady electric field is directed toward the center of the sheet and produces a shear flow $v_0(x)$ $(=E_0(x)/B_0)$ having a value at $x=\pm a$ of

$$v_0(a) \equiv v_0, \; v_0(-a) \equiv -v_0. \tag{3.60}$$

Let us now consider the surface wave created by the surface charge due to an undulating boundary at $x=\pm a$. The surface wave is generated only by the charges at the surface. The electrostatic-field equation is thus the Laplace equation for the electrostatic potential φ_1,

$$\nabla^2 \varphi_1 = 0, \tag{3.61}$$

meaning that the wave that we are considering is an incompressible mode. Eq. (3.61) is identical to the MHD case, Eq. (3.52) and has a general solution for a two-dimensional system such as that shown in Fig. 28, given by

$$\varphi_1 \sim e^{i(ky-\omega t)} \cdot e^{\pm kx}. \tag{3.62}$$

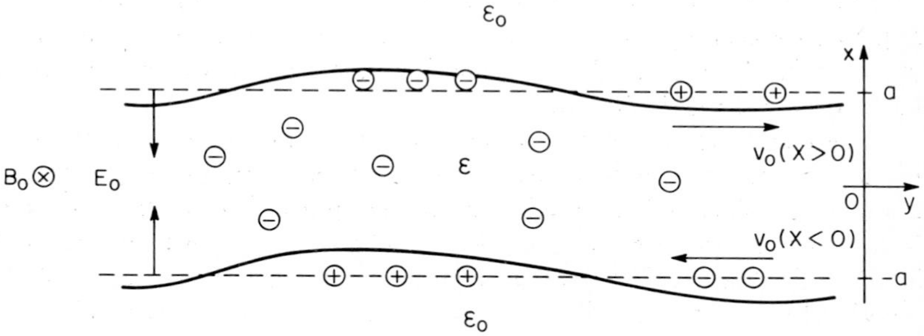

Fig. 28. Negatively charged sheet in a magnetic field B_0. The nonuniform electric field in the sheet generates a shear flow $\pm v_0$ at $x=\pm a$. A perturbation at the surface gives rise to a surface charge that leads to the Kelvin-Helmholtz instability

Hence we choose

$$\begin{aligned} \varphi_1(x>a) &= A\,e^{i(ky-\omega t)}\,e^{-kx} \\ \varphi_1(-a<x<a) &= e^{i(ky-\omega t)}(B\,e^{kx}+C\,e^{-kx}) \\ \varphi_1(x<-a) &= D\,e^{i(ky-\omega t)}\,e^{kx}. \end{aligned} \tag{3.63}$$

The boundary conditions at $x=\pm a$ are (1) the continuity of the tangential electric field $E_y=-\partial\varphi_1/\partial y$ and (2) a discontinuity of the normal electric field by an amount equal to the surface charge ϱ_s, i.e.

$$\begin{aligned} \left.\frac{\partial\varphi_1}{\partial x}\right|_{x\to a^+} - \left.\frac{\partial\varphi_1}{\partial x}\right|_{x\to a^-} &= -\frac{\varrho_s(a)}{\varepsilon_0} \\ \left.\frac{\partial\varphi_1}{\partial x}\right|_{x\to -a^+} - \left.\frac{\partial\varphi_1}{\partial x}\right|_{x\to -a^-} &= -\frac{\varrho_s(-a)}{\varepsilon_0} \end{aligned} \tag{3.64}$$

where ϱ_s can be obtained from the equation of continuity

$$\frac{\partial n_1}{\partial t}+\nabla\cdot(n_1\boldsymbol{v}_0+n_0\boldsymbol{v}_1)=0. \tag{3.65}$$

Following Buneman *et al.* (1966), we consider here a low-frequency perturbation such that $\omega\ll\omega_{ce}$. Then the perturbed velocity $\boldsymbol{v}_1$ in Eq. (3.65) can be expressed simply by

$$\boldsymbol{v}_1=\frac{\boldsymbol{E}_1\times\boldsymbol{B}_0}{B_0^2}. \tag{3.66}$$

We are considering an electrostatic perturbation, $\nabla\times\boldsymbol{E}_1=0$, hence from Eq. (3.66), $\nabla\cdot\boldsymbol{v}_1=0$. Eq. (3.65) then can be reduced to

$$\frac{\partial n_1}{\partial t}+\boldsymbol{v}_0\cdot\nabla n_1=-\boldsymbol{v}_1\cdot\nabla n_0. \tag{3.67}$$

By substituting Eq. (3.66) into Eq. (3.67) we obtain the number density perturbation n_1

$$n_1(\pm a)=\frac{k\varphi_1}{\omega\mp k v_0}\,\frac{1}{B_0}\,\frac{\partial n_0}{\partial x}. \tag{3.68}$$

Now, if the charged sheet has a sharp boundary at $x = \pm a$ as assumed, $n_0(x)$ has the form of a unit step function $U(x)$, which may be written

$$n_0(x) = n_0 [U(x+a) - U(x-a)] \tag{3.69}$$

and

$$\frac{\partial n_0}{\partial x} = n_0 [\delta(x+a) - \delta(x-a)]. \tag{3.70}$$

Hence, the surface charge density ϱ_s can be obtained from Eqs. (3.68) and (3.70)

$$\varrho_s(\pm a) = \pm \frac{e n_0}{B_0} \frac{k \varphi_1}{\omega \mp k v_0}. \tag{3.71}$$

The dispersion relation can be obtained by applying the continuity of φ_1 and the boundary conditions given by Eq. (3.64) to the solution of the Laplace equation (3.63);

$$\frac{4\omega^2}{\omega_0^2} = \left(1 - \frac{2 k v_0}{\omega_0}\right)^2 - e^{-4ka} \tag{3.72}$$

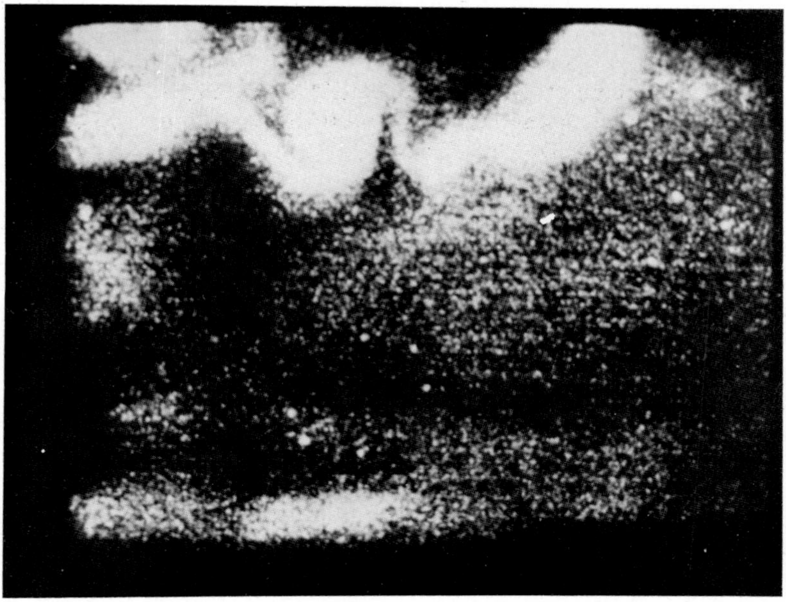

Fig. 29. Observation of Kelvin-Helmholtz instability in an auroral sheet. (After Hallinan and Davis, 1971)

where

$$\omega_0 = \frac{e n_0}{\varepsilon_0 B_0} = \frac{\omega_{pe}^2}{\omega_{ce}}.$$

The dispersion relation derived here is identical to the one for an incompressible fluid (Chandrasekhar, 1961). The instability occurs when the right-hand side of Eq. (3.72) is negative, or for a wave number k satisfying

$$2ka \lesssim 1.3. \tag{3.73}$$

Therefore, the instability occurs when the wavelength in the direction of the shear flow is comparable to or longer than 2π times the width of the charged sheet $2a$. A consequence of the instability is the deformation of the sheet into periodic curls around the magnetic lines of force. Hallinan and Davis (1970) have applied this instability to the formation of curls in the auroral sheet. Since the instability of a charged sheet grows for a wave propagating in the direction of the shear flow, for an electron sheet the curls produced by the instability will be in the clockwise direction looking in the direction of the magnetic field. On the other hand, for a sheet of positive ions, although the direction of the shear flow is given by the same formula, $\boldsymbol{E} \times \boldsymbol{B}$, the direction of $\boldsymbol{E}$ is reversed; hence so is the direction of the curls. For either case, the curls are formed in the direction of the cyclotron motion of the particles (cf. Fig. 29).

3.5 Current Pinch Instabilities

3.5a Instability of Cylindrical Current with Infinite Conductivity

In this subsection, we consider instabilities driven by a current in a plasma of finite cross section. As we have seen, a current can produce a two-stream instability even in a uniform plasma. In that case, the threshold velocity of the electrons is given by the ion sound velocity. For a current in a plasma with a finite cross section, extra free energy is available from the nonuniformity in space; therefore the threshold of the instability can become lower.

First let us consider a plasma with a circular cross section. Three kinds of deformation of such a circular pinch are considered in Fig. 30. The first case, case a, is called sausage-type instability. The azimuthal magnetic field $B_{0\theta}$ produced by the current I_0 becomes stronger at the neck point because $B_{0\theta} \sim \mu_0 I_0 / 2\pi r$; hence the perturbation tends to

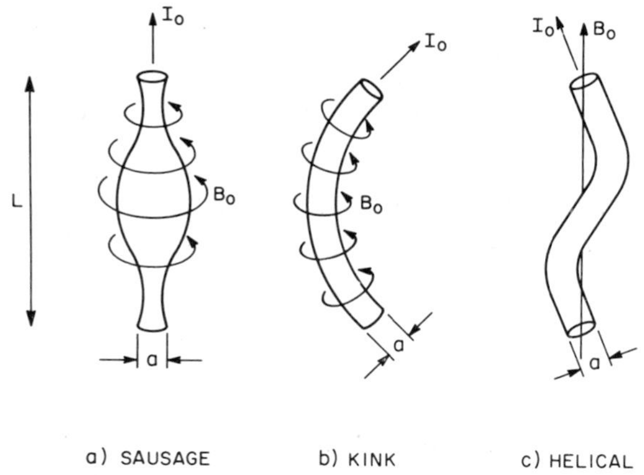

Fig. 30a–c. Various shapes of instabilities of cylindrical current pinches

grow. The threshold condition for the instability is obtained as follows. For a shear mode, plasma moves with the magnetic field; hence the total flux inside the plasma is constant

$$r^2 B_{0z} = \text{const} \tag{3.74}$$

where B_{0z} is the axial magnetic field. The change in B_{0z} associated with the change in radius r of the plasma column is then given by

$$\delta B_{0z} = -B_{0z} \frac{2dr}{r}. \tag{3.75}$$

On the other hand, the associated change in the azimuthal magnetic field $B_{0\theta}$ is given by

$$\delta B_{0\theta} = -B_{0\theta} \frac{dr}{r} \tag{3.76}$$

because the current I_0 is constant. The total change of the magnetic-field pressure directed inward is then

$$\delta p = \delta\left(\frac{B_{0\theta}^2}{2\mu_0} - \frac{B_{0z}^2}{2\mu_0}\right) = -\left(\frac{B_{0\theta}^2}{\mu_0} - \frac{2B_{0z}^2}{\mu_0}\right)\frac{dr}{r}. \tag{3.77}$$

The instability condition is simply that the change in the magnetic pressure associated with an increase of radius dr is negative. Thus the pinch is unstable against the sausage-type perturbation when

$$B_{0\theta}^2 > 2\,B_{0z}^2\,. \tag{3.78}$$

The stability condition for the kink-type perturbation shown in Fig. 30b can be obtained in the same way, and the condition of the instability for such a case becomes

$$B_{0\theta}^2 \ln\left(\frac{L}{a}\right) > B_{0z}^2 \tag{3.79}$$

where L and a are the length and the radius, respectively, of the current pinch.

In the magnetosphere, field-aligned currents have often been observed during substorm times (e.g., Zmuda *et al.*, 1966; Cummings and Dessler, 1967; Cloutier *et al.*, 1970). However, the maximum horizontal field $B_{0\theta}$ produced by these currents is of the order of $10^3\ \gamma\ (\gamma = 10^{-5}$ gauss), which is less than one-tenth of the geomagnetic field B_{0z}. Therefore, instabilities of types a and b discussed above are unlikely to occur.

However, the instability of type c in Fig. 30, the helical type (also called kink instability) is quite likely to occur because of its much lower threshold. The threshold condition of the helical-type instability will be shown to be

$$B_{0\theta} > \frac{2\pi a}{L}\, B_{0z}\,. \tag{3.80a}$$

While instabilities of types a and b are primarily due to the pinch (compressional) effect of the current-generated magnetic field, the type c instability is due to the tension of the field lines that are bent into a helical shape by the current. While the tension of the bent field lines causes them to tend to straighten, the current tends to deform into a helical shape.

Let us derive here the condition of the instability, Eq. (3.80a), following Kadomtsev (1966). We assume that the plasma is collisionless and hence has an infinite conductivity. In such a case the current flows only at the surface of the column. Furthermore, we assume an incompressible perturbation.

Because the current flows only at the surface, the unperturbed magnetic field inside the plasma column $\boldsymbol{B}_0^i$ is uniform and $\boldsymbol{B}_0^i = B_0^i\,\boldsymbol{e}_z$. Then, from Eq. (3.50), we have the same wave equation as given by Eq. (3.52),

$$\nabla^2 \tilde{p}_1 = 0 \tag{3.81}$$

where $\tilde{p}_1$ is the perturbed total pressure given by

$$\tilde{p}_1 = p_1 + \frac{\boldsymbol{B}_1^i \cdot \boldsymbol{B}_0^i}{\mu_0}. \tag{3.82}$$

In the same way, the relation between the plasma displacement ξ and the total pressure $\tilde{p}_1$ is obtained from Eq. (3.57)

$$\xi = \frac{\nabla \tilde{p}_1}{m_i n_0 (\omega^2 - k^2 v_A^2)} \tag{3.83}$$

where $v_A = B_0^i/(\mu_0 m_i n_0)^{1/2}$ and k is the wave number in the axial (z) direction.

Outside the plasma, we have $\nabla \times \boldsymbol{B}_1^e = 0$, as well as $\nabla \cdot \boldsymbol{B}_1^e = 0$, hence $\boldsymbol{B}_1^e$ can be expressed as a gradient of a scalar function ψ,

$$\boldsymbol{B}_1^e = \nabla \psi \tag{3.84}$$

where

$$\nabla^2 \psi = 0. \tag{3.85}$$

Thus we have obtained the wave equations both inside (3.81) and outside (3.85) the plasma. We need boundary conditions to connect solutions of these wave equations. Compared to the case of the Kelvin-Helmholtz instability, the boundary conditions for this case are more complex because of the plasma-vacuum contact and also because the unperturbed magnetic field outside the plasma is nonuniform. The first boundary condition is the condition of the pressure balance:

$$p_0 + p_1 + \frac{1}{2\mu_0}\left(\boldsymbol{B}_0^i + \boldsymbol{B}_1^i\right)^2 = \frac{1}{2\mu_0}\left(\boldsymbol{B}_0^e + \boldsymbol{B}_1^e\right)^2. \tag{3.86}$$

To obtain the boundary condition for the perturbed quantities, we evaluate this expression at a displaced boundary at $\boldsymbol{r} = \boldsymbol{r}_0 + \boldsymbol{\xi} = \boldsymbol{r}_0 + \xi_n \boldsymbol{n}$, where $\boldsymbol{n}$ is the unit normal vector at the surface and $\xi_n (= \xi_r)$ is the normal displacement. Expanding Eq. (3.86) into powers of the perturbed quantities and retaining the linear terms, we obtain:

$$\tilde{p}_1 = \frac{\boldsymbol{B}_0^e \cdot \boldsymbol{B}_1^e}{\mu_0} + \frac{\xi_n}{2\mu_0}\left[\frac{\partial (B_0^e)^2}{\partial n} - \frac{\partial (B_0^i)^2}{\partial n}\right], \tag{3.87a}$$

or for a cylindrical geometry, noting that that B_0^i is constant,

$$\tilde{p}_1 = \frac{\boldsymbol{B}_0^e \cdot \boldsymbol{B}_1^e}{\mu_0} + \frac{\xi_r}{2\mu_0}\frac{\partial (B_0^e)^2}{\partial r}. \tag{3.87b}$$

The second boundary condition is the continuity of the tangential electric field. Because of the assumed infinite conductivity of the plasma, the tangential component of the external electric field should be zero, that is,

$$\boldsymbol{E}_{1t} + (\boldsymbol{v}_1 \times \boldsymbol{B}_0^e)_t = 0 \tag{3.88a}$$

where subscript t indicates the tangential components. Using Maxwell's equation $\nabla \times \boldsymbol{E}_1 = -\partial \boldsymbol{B}_1/\partial t$ and the displacement vector ξ, this expression reduces to

$$\boldsymbol{n} \cdot \boldsymbol{B}_1^e = \boldsymbol{n} \cdot \nabla \times (\xi \times \boldsymbol{B}_0^e). \tag{3.88b}$$

We now have all the necessary boundary conditions and the wave equations for both the internal and external regions of the plasma. Let us then solve the wave equations for the cylindrical geometry. From Eq. (3.81), we have for the perturbed total pressure of the plasma:

$$\tilde{p}_1 = A \frac{I_n(kr)}{I_n(ka)} e^{i(n\theta + kz)} \tag{3.89}$$

where I_n is the modified Bessel function of the first kind and A is an integration constant designating the value of $\tilde{p}_1$ at $r = a$. The corresponding radial displacement of the plasma $\xi_r(r)$ can be obtained from Eqs. (3.83) and (3.89)

$$\xi_r(r) = \frac{k}{m_i n_0 (\omega^2 - k^2 v_A^2)} A \frac{I_n'(kr)}{I_n(ka)}. \tag{3.90}$$

For outside the plasma, we have the wave equation given by Eq. (3.85). The solution of ψ bound as $r \to \infty$ is

$$\psi = \frac{C K_n(kr)}{K_n(ka)} \tag{3.91}$$

where K_n is the modified Bessel function of the second kind and C is the integration constant.

Now we use the boundary conditions. First we take the pressure balance condition, Eq. (3.87b). Outside the plasma, the unperturbed magnetic field $\boldsymbol{B}_0^e$ has axial and azimuthal components B_{0z}^e and $B_{0\theta}^e$. Since B_{0z}^e is assumed uniform, while $B_{0\theta}^e \sim 1/r$,

$$\frac{\partial}{\partial r} [(B_{0z}^e)^2 + (B_{0\theta}^e)^2] = -2 [B_{0\theta}^e(a)]^2/a$$

at the boundary $r=a$. Then Eq. (3.87b) gives

$$A=\frac{i}{\mu_0}\left(k B_{0z}^e+\frac{\mathrm{n}}{a} B_{0\theta}^e\right) C-\frac{(B_{0\theta}^e)^2}{\mu_0 a} \xi_r(a). \tag{3.92}$$

In a similar way, the other boundary condition, that of the vanishing tangential electric field given by Eq. (3.88b) gives

$$i\left(k B_{0z}^e+\frac{\mathrm{n}}{a} B_{0\theta}^e\right) \xi_r(a)=C k \frac{K_{\mathrm{n}}'(k a)}{K_{\mathrm{n}}(k a)}. \tag{3.93}$$

Combining Eqs. (3.90), (3.92) and (3.93) and eliminating $\xi_r(a)$, A, and C, we obtain the following dispersion relation

$$\mu_0 m_i n_0 \omega^2=k^2 B_0^2-\left(k B_{0z}^e+\frac{\mathrm{n}}{a} B_{0\theta}^e\right)^2 \frac{I_{\mathrm{n}}'(k a) K_{\mathrm{n}}(k a)}{I_{\mathrm{n}}(k a) K_{\mathrm{n}}'(k a)}-\frac{(B_{0\theta}^e)^2 k}{a} \frac{I_{\mathrm{n}}'(k a)}{I_{\mathrm{n}}(k a)}. \tag{3.94}$$

For instability, the right-hand side has to give a negative value. The first and the second terms are positive because $K_{\mathrm{n}}/K_{\mathrm{n}}'<0$, while $I_{\mathrm{n}}'/I_{\mathrm{n}}>0$. Hence it is the last term on the right-hand side that gives rise to an unstable solution. This negative contribution originates from the fact that $(\partial/\partial r)(B_{0e}^2)<0$; that is, the external magnetic field pressure decreases with radial displacement of the plasma column.

For field-aligned currents in the magnetosphere, we can assume $B_{0z}^e \gg B_{0\theta}^e$. In this case, a long wavelength perturbation characterized by $ka \ll 1$ can lead to an unstable solution. At small ka, $I_{\mathrm{n}}'/I_{\mathrm{n}}=\mathrm{n}/ka$, $K_{\mathrm{n}}'/K_{\mathrm{n}}=-\mathrm{n}/ka$; thus the dispersion relation reduces to

$$\mu_0 m_i n_0 \omega^2=k^2 B_0^2+\left(k B_{0z}^e+\frac{\mathrm{n}}{a} B_{0\theta}^e\right)^2-\frac{\mathrm{n}(B_{0\theta}^e)^2}{a^2}. \tag{3.95}$$

One can see from Eq. (3.95) that a perturbation with $k<0$ and for an azimuthal mode with $\mathrm{n}=1$ is unstable (helical perturbation) and the condition of instability can be found to be $|k|<B_{0\theta}^e/a B_{0z}^e$ or

$$\frac{B_{0\theta}^e}{B_{0z}^e}>\frac{2\pi a}{L}, \tag{3.80a}$$

the form presented before. This instability was derived by Kruskal and Schwarzschild (1954) and Shafranov (1956) and the axial current corresponding to $B^e_{0\theta}(=\mu_0 I_0/2\pi a)$ in this expression is called the Kruskal-Shafranov limit. Let us investigate the applicability of this expression to the field aligned currents during auroral breakup. Here, if we consider a current flowing along the field line from the equator to the ground, the cross section of the current as well as the unperturbed flux density B_{0z} changes along the current path. However, because $B_{0\theta}=\mu_0 I_0/2\pi a$ and $\pi a^2 B_{0z}=\text{const}$, we can see that the instability condition (3.80a) does not depend on position along the current path, and hence can be evaluated at any point. If we take as typical values near the ground, $B_{0z}\sim 5\times 10^4\,\gamma$, $L\sim 1.5\times 5\ R_E=4.7\times 10^4$ km and $B_{0z}\sim 500\ \gamma$, the circumferential length of the current cross section $2\pi a$ becomes smaller than 4.7×10^2 km. That is, if the current flows down in the ionosphere with a cross sectional radius of approximately 100 km, the helical instability occurs, and the current path is bent.

In terms of the current density J, the instability condition can be expressed as

$$J=\frac{I_0}{\pi a^2}>\frac{4\pi B_{0z}}{\mu_0 L}. \tag{3.80b}$$

Obviously the critical current density is a function of the location along the field line. If we again take the example near the ground, the critical current density there becomes $\sim 10^{-5}$ ampere/m^2.

3.5b Instability of a Current with Finite Resistivity: Tearing Mode Instability

We now shift our interest to the pinch of an *infinitely extended sheet* current. In this case none of the instabilities which appeared for cylindrical pinches is expected to occur because all of those instabilities originated from the radially decreasing magnetic field pressure, whereas a sheet current generates an external magnetic field which remains constant.

In such a geometry, an instability occurs only in the presence of a finite resistivity in the plasma. This resistivity works to dissipate the current collectively (and thus the current-generated magnetic field); hence the instability is called the tearing mode instability. The instability may occur in the presence or absence of a stationary magnetic field in the direction of the current. We consider here the case without such a magnetic field: the entire magnetic field here is produced by the sheet current. We consider this problem in relation to the instability of the neutral sheet in the magnetospheric tail.

That a neutral sheet is subject to instability was first pointed out by Dungey (1958) and elaborated later by Furth *et al.* (1963). Consider a sheet current that is infinitely extended in the yz plane and flowing in

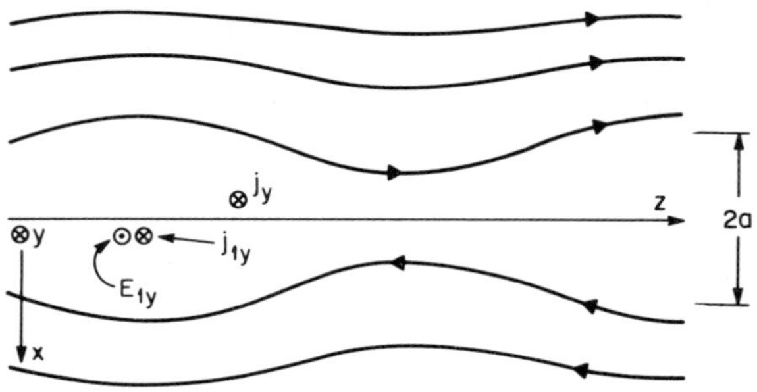

Fig. 31. Coordinate system used in the derivation of the instability condition for a sheet current

the y direction as shown in Fig. 31. Such a current is sandwiched by a self-generated magnetic field in the z direction that pinches the current to an equilibrium size. Perturbations in the current $\boldsymbol{J}_1$ and the magnetic field $\boldsymbol{B}_1$ can be shown to produce a stably propagating wave if the plasma is assumed to be a perfect conductor. Only when the plasma has a finite resistivity does an instability occur. In this sense, the mechanism of the instability differs considerably from the instabilities of a cylindrical pinch. To show how the finite resistivity produces the instability, we use Eq. (1.27) keeping the effect of finite resistivity, i.e.

$$\boldsymbol{E}_1 + \boldsymbol{v}_1 \times \boldsymbol{B}_0 - \eta \boldsymbol{J}_1 = 0 \tag{3.96}$$

where η is the plasma resistivity in ohm-meters. From this expression we can see that the effect of finite resistivity becomes important at the neutral layer at $x \sim 0$, where the z-directed magnetic field $B_0 \sim 0$. On the other hand, at distances sufficiently far from the neutral layer, the $\boldsymbol{v} \times \boldsymbol{B}$ term can dominate, and the plasma can be regarded as lossless. To understand the physical process of the instability, we hence choose a simple model in which the current layer is divided into two regions, that of resistivity at $|x| < \varepsilon$, and that of no resistivity at $a > |x| > \varepsilon$.

Let us first consider the dynamics in the resistive region, $|x| < \varepsilon$. In this region $\boldsymbol{E}_1$ may be expressed as $\eta \boldsymbol{J}_1$ from Eq. (3.96). Then, using Maxwell's equation, we have

$$\frac{\partial \boldsymbol{B}_1}{\partial t} = \frac{\eta}{\mu_0} \nabla^2 \boldsymbol{B}_1 . \tag{3.97}$$

Eq. (3.97) represents simply the skin effect of the plasma. For an eigenfunction of the form e^{ikx}, Eq. (3.97) gives a solution with a negative imaginary part of ω, indicating dissipation of wave energy and no instability. However, if the field solution at $|x|>\varepsilon$ allows a solution, through boundary conditions, such that $\boldsymbol{B}_1 \sim e^{\pm \kappa x}$ at $|x|<\varepsilon$, a positive imaginary ω solution results. Anticipating such a case, we put, say for B_{1x} components, $B_{1x} \sim B_{1x}(x) e^{ikz+\gamma t}$, and Eq. (3.97) becomes

$$\frac{d^2 B_{1x}}{dx^2} - \left(k^2 + \frac{\gamma \mu_0}{\eta}\right) B_{1x} = 0 \tag{3.98}$$

which can immediately be solved to give*

$$B_{1x} \sim A \cosh \left(k^2 + \frac{\gamma \mu_0}{\eta}\right)^{1/2} x. \tag{3.99}$$

We now consider the lossless region. From Eq. (3.96) and the Maxwell's equation (1.15), we can express the x component of the velocity perturbation by B_{1x} as

$$i k v_{1x} B_0 = \gamma B_{1x} \tag{3.100}$$

where $B_0 = B_0(x)$ is the dc magnetic field produced by the sheet current.

Another equation that relates v_{1x} and B_{1x} can be obtained from the MHD equation (1.25) and Maxwell's equation (1.30); the pressure-gradient term is eliminated by taking the curl of Eq. (1.25). Assuming incompressibility, $\nabla \cdot \boldsymbol{v}_1 = 0$, and using $\nabla \cdot \boldsymbol{B} = 0$, we derive the following wave equation

$$\frac{d^2 B_{1x}}{dx^2} - \left(k^2 + \frac{B_0''}{B_0}\right) B_{1x} = \frac{\gamma^2}{k^2 v_A^2} \left(\frac{d^2}{dx^2} - k^2\right) \left(\frac{B_{1x}}{B_0}\right). \tag{3.101 a}$$

This can be reduced further for a small growth rate $\gamma^2 \ll k^2 v_A^2$, to

$$\frac{d^2 B_{1x}}{dx^2} - \left(k^2 + \frac{B_0''}{B_0}\right) B_{1x} = 0, \tag{3.101 b}$$

where B_0'' is the second derivative of B_0 with respect to x. If B_0 is uniform so that $B_0''=0$, Eq. (3.101) simply shows the electromagnetic cutoff mode in space. However, for a nonuniform sheet current confined within $|x| \le a$, B_0''/B_0 can be seen to become negative as shown in Fig. 32. Eq. (3.101) then admits a sinusoidal solution for small k. For example, if

* There exists an alternative solution to this, which is an odd function of x, i.e., $\sinh(k^2 + \gamma \mu_0/\eta)^{1/2} x$. However, it can be shown that this solution does not satisfy the matching boundary condition at $x = \pm \varepsilon$.

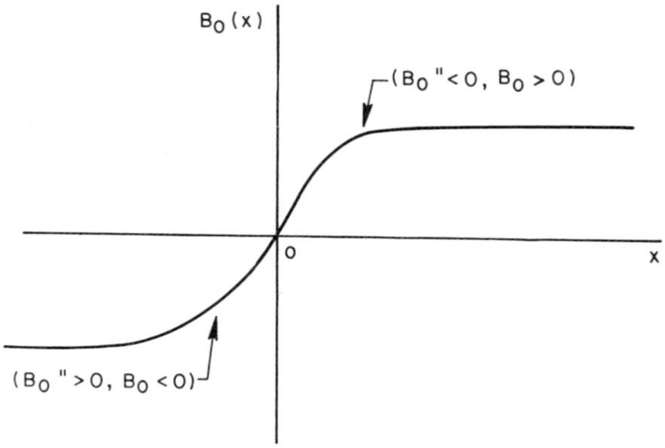

Fig. 32. Variation of the z directed magnetic flux density $B_0(x)$ associated with the sheet current, as a function of x

we write $B_0''/B_0 \sim -\lambda^{-2}$ the equation becomes

$$\frac{d^2 B_{1x}}{dx^2} + \left(\frac{1}{\lambda^2} - k^2\right) B_{1x} = 0 \tag{3.102}$$

or

$$B_{1x} = C \sin\left(\frac{1}{\lambda^2} - k^2\right)^{1/2} x. \tag{3.103}$$

If we now connect this solution at $x = \varepsilon$ to the B_{1x} obtained for $|x| < \varepsilon$ in Eq. (3.99), we can derive the growth rate γ as

$$\gamma = \eta/(\varepsilon^2 \mu_0). \tag{3.104}$$

Although we cannot obtain an exact value of the growth rate from the above expression because ε is a quantity assumed in the derivation, we can understand the mechanism of the instability from the above argument. The driving force of the instability is the nonuniform magnetic field with $B_0/B_0'' < 0$. The instability occurs for a wave length in the z direction longer than the thickness, i.e., $k < 1/\lambda \sim 1/a$, and for a plasma with a finite resistivity η. As a consequence of the instability, x-type neutral points tear the sheet current into a number of smaller segments (Fig. 33).

The magnetospheric tail can hardly be considered as "resistive"; therefore one might conclude that the tearing mode instability is not applicable there. However, if one takes into account the interactions between waves and warm particles, the tearing mode may become possible (Coppi *et al.*, 1966; Hoh, 1966). Hoh has shown that electron Landau damping can contribute to the resistivity, and that the instability

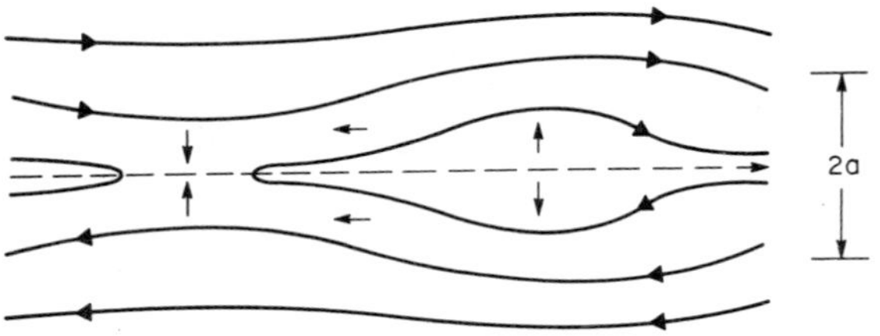

Fig. 33. Tearing-mode instability of the neutral sheet

can occur in the absence of collisional resistivity. Coppi *et al.* (1966), using the same idea, have explicitly calculated the growth time (time needed to exponentiate the perturbation) using parameters in the actual magnetospheric tail and have shown it to be of the order of 10 seconds.

However, two difficulties in this explanation have been pointed out by Laval and Pellat (1968) and Hoh and Bers (1966). According to Laval and Pellat, the instability is stabilized when the anisotropy in temperature is taken into account. That is, the collisionless tearing mode is stable when

$$1-\frac{T_\perp}{T_\parallel}>\frac{\varrho_e}{a} \tag{3.105}$$

where ϱ_e is the electron cyclotron radius. Because $\varrho_e/a \ll 1$, Eq. (3.105) implies that the mode is stable essentially when $T_\perp < T_\parallel$. Hoh and Bers have calculated the number of electrons that contribute to the resonant interaction through Landau damping and, by considering the resultant source of free energy, they have obtained the maximum amplitude of the field perturbations. For a reasonable choice of parameters, in the geomagnetic tail, they have shown that the maximum amplitude of the magnetic field perturbation produced by the instability must be of the order of or smaller than 0.5γ, which is much smaller than the average tail field ($\sim 15\gamma$).

Later Biskamp *et al.* (1970) in fact showed that when the tearing mode instability is excited in a collisionless plasma, the excited turbulence increases $T_\parallel$ through quasilinear diffusion and the plasma is stabilized by reaching the condition (3.105).

In view of these arguments, the use of the collisionless tearing-mode instability to describe gross tail dynamics, or to cause magnetic substorms, seems rather difficult. The possibility that the magnetic substorm may be caused by the tearing mode instability can be raised when the intensity of the neutral sheet current is large enough to cause a two stream type microinstability. This may generate enough anomalous resistivity to provide a sufficient amount of dissipation for the tearing mode instability.

Because of the general presence of a northward magnetic field, a possible candidate for such a microinstability is the beam cyclotron instability briefly discussed in the end of the Subsection 2.2a.

References for Chapter 3

Biskamp, D., Sagdeev, R.Z., Schindler, K.: Nonlinear evolution of the tearing instability in the geomagnetic tail. Cosmic Electrodynamics **1**, 297 (1970)

Boller, B.R., Stolov, H.L.: Kelvin-Helmholtz instability and semiannual variation of geomagnetic activity. J. Geophys. Res. **75**, 6073 (1970)

Buneman, O., Levy, R.H., Linson, L.M.: Stability of crossfield electron beams. J. Appl. Phys. **37**, 3203 (1966)

Chandrasekhar, S.: Hydrodynamic and hydromagnetic stability, chapt. 13. Oxford: Clarendon, 1961

Chang, D.B., Pearlstein, L.D., Rosenbluth, M.N.: On the interchange instability of the Van Allen belt. J. Geophys. Res. **70**, 3085 (1965)

Chang, D.B., Pearlstein, L.D., Rosenbluth, M.N.: Corrections and additions to the paper entitled: On the interchange instability of the Van Allen belt. J. Geophys. Res. **71**, 351 (1966)

Chen, L., Hasegawa, A.: Theory of magnetic pulsations I. Steady state excitation of field line resonance. J. Geophys. Res. **79**, 1024 (1974)

Cloutier, P.A., Anderson, H.R., Park, R.J., Vondrak, R.R., Spiger, R.J., Sandel, B.R.: Detection of geomagnetically aligned currents associated with an auroral arc. J. Geophys. Res. **75**, 2595 (1970)

Coppi, B., Laval, G., Pellat, R.: Dynamics of geomagnetic tails. Phys. Rev. Letters **16**, 1207 (1966).

Cummings, W.D., Dessler, A.J.: Field-aligned currents in the magnetosphere. J. Geophys. Res. **72**, 1007 (1967)

Dungey, J.W.: Cosmic electrodynamics, p. 98. New York: Cambridge University Press, 1958

Eviatar, A., Wolf, R.A.: Transfer processes in the magnetosphere. J. Geophys. Res. **73**, 5161 (1968)

Furth, H.P., Killeen, J., Rosenbluth, M.N.: Finite resistivity instabilities of a sheet pinch. Phys. Fluids **6**, 459 (1963)

Hagege, K., Laval, G., Pellat, R.: Interaction between high frequency turbulence and magnetospheric micropulsations. J. Geophys. Res. **78**, 3806 (1973)

Hallinan, T.J., Davis, T.N.: Small scale auroral arc distortions. Planetary Space Sci. **18**, 1735 (1970)

Hasegawa, A.: Drift-wave instabilities of a compressional mode in a high β plasma. Phys. Rev. Letters **27**, 11 (1971).

Hoh, F.C.: Stability of sheet pinch. Phys. Fluids **9**, 277 (1966)

Hoh, F.C., Bers, A.: Resonant particle energy transfer to magnetic perturbations with application to the geomagnetic tail. Boeing Sci. Res. Lab. Doc. D1-82-0539 (1966)

Kadomtsev, B.B.: Reviews of plasma physics, ed. by M.A. Leontovich, p. 174. New York: Consultants Bureau, 1966

Krall, N.A.: Advances in plasma physics, vol. 1, ed. by A. Simon and W.B. Thompson, p. 153. New York: Interscience, 1968

Kruskal, M., Schwarzschild, M.: Some instabilities of a completely ionized plasma. Proc. Roy. Soc. London, Ser. A **223**, 348 (1954)

Laval, G., Pellat, R.: Stability of the plane neutral sheet for oblique propagation and anisotropic temperature. Proc. ESRIN Study Group Frascati (Rome), Italy, December 1967, ESRO SP-36 (July 1968)

Lehnert, B.: Stability of a plasma boundary in a magnetic field. Phys. Fluids **4**, 847 (1961)

Lerche, I.: On the boundary layer between a warm streaming plasma and a confined magnetic field. J. Geophys. Res. **72**, 5295 (1967)

Mikhailovskaya, L.V., Mikhailovskii, A.B.: Drift instability in a dense plasma. Zh. Eksp. Teor. Fiz. **45**, 1566 (1963) [Soviet Phys. JETP Engl. Transl. **18**, 1077 (1964)]

Mikhailovskii, A.B.: Review of plasma physics, ed. by M.A. Leontovich, p. 172, New York: Consultant Bureau, 1967

Mikhailovskii, A.B., Timofeev, A.V.: Theory of cyclotron instability in a non-uniform plasma. Zh. Eksp. Teor, Fiz. **44**, 919 (1963) [Soviet Phys. JETP Engl. Transl. **17**, 626 (1963)]

Ong, R.S.B., Roderick, N.: On the Kelvin-Helmholtz instability of the earth's magnetopause. Planetary Space Sci. **20**, 1 (1972)

Parker, E.N.: Confinement of a magnetic field by a beam of ions. J. Geophys. Res. **72**, 2315 (1967a)

Parker, E.N.: Small-scale nonequilibrium of the magnetopause and its consequences. J. Geophys. Res. **72**, 4365 (1967b)

Parker, E.N.: Reply to comment by H.E. Stubbs. J. Geophys. Res. **73**, 2540 (1968a)

Parker, E.N.: Reply to comment by V.C.A. Ferraro and C.M. Davis. J. Geophys. Res. **73**, 3607 (1968b)

Rosenbluth, M.N., Longmire, C.L.: Stability of plasmas confined by magnetic field. Ann. Phys. (N.Y.) **1**, 120 (1957)

Rudakov, L.I., Sagdeev, R.Z.: On the instability of a nonuniform rarefield plasma in a strong magnetic field. Dokl. Akad. Nauk SSSR **138**, 581 (1961) [Soviet Phys. "Doklady" English Transl. **6**, 415 (1961)]

Sen, A.K.: Stability of hydromagnetic Kelvin-Helmholtz discontinuity. Phys. Fluids **6**, 1154 (1963)

Sen, A.K.: Effect of compressibility on Kelvin-Helmholtz instability in a plasma. Phys. Fluids **7**, 1293 (1964)

Shafranov, V.D.: The stability of a cylindrical gaseous conductor in a magnetic field. Atomnaya energiya **5**, 38 (1956) [Soviet J. At. Energy English Transl. **1**, 709 (1956)].

Southwood, D.J.: The hydromagnetic stability of the magnetospheric boundary. Planetary Space Sci. **16**, 587 (1968).

Zmuda, A., Martin, J.H., Heuring, F.T.: Transverse magnetic disturbances at 1100 kilometers in the auroral region. J. Geophys. Res. **71**, 5033 (1966)

4. Nonlinear Effects Associated with Plasma Instabilities

4.1a Introduction

When plasma instabilities are excited and waves grow, a number of interesting nonlinear phenomena appear which modify the plasma states. Ordinarily, the nonlinear effects become important when the wave energy density W becomes much larger than at the thermal equilibrium: $W \gg n_0 T(n_0 \lambda_D^3)^{-1}$, but is yet much smaller than the thermal energy density: $W \ll n_0 T$. The ratio of the wave energy density and the thermal energy density, $(W/n_0 T)$, is used as a measure of turbulence.

In this chapter we introduce the nonlinear effects which appear as a consequence of the various plasma instabilities discussed in the foregoing chapters: in particular those which result from the microinstabilities discussed in Chapter 2. We will present here only philosophical aspects of the various nonlinear phenomena and will not go into the details and particular applications, because even for a given instability, dominating nonlinear effects vary very much depending on actual circumstances.

As we have seen, an instability is a process in which free energy in the plasma is exponentially converted into fluctuating electromagnetic field energy. As the field energy grows, a nonlinearity of the plasma may cause a change in some quantity which may or may not be directly involved with the instability, but which can cause the instability to saturate. Two types of effects may result: effects on the plasma particles and effects on the waves.

In Section 4.2, we discuss effects on particles, including quasilinear diffusion of a distribution function, resonance broadening, and particle trapping by waves. In Section 4.3, we discuss effects on waves, including nonlinear wave interactions (the decay instability) and wave-particle interactions (nonlinear Landau damping). In Section 4.4, we discuss some coherent nonlinear wave phenomena, such as the formation of shock and solitary waves, modulational instabilities and the formation of envelope solitons.

4.1b Methods of Approach

To analyze the nonlinear development of an instability, one ordinarily adopts a perturbation technique, in which one assumes that at an early stage of development the system has a response that obeys the linear dispersion relation and that at a later time, as the amplitude grows larger, the nonlinearity modifies the linear properties at a rate much slower than the linear response of the system. Such modifications occur in general at a rate $\tau \sim (W/n_0 T)^{-1} t_L$, where t_L is the linear time scale ($t_L \sim \omega^{-1}$), and it occurs both on linearly unperturbed quantities such as the background distribution function f_0 and/or on linearly perturbed quantities such as the wave amplitude. Similarly the space dependency of these quantities changes on a scale $\boldsymbol{\xi} \sim (W/n_0 T)^{-1} \boldsymbol{x}_L$, where $\boldsymbol{x}_L$ is the linear spatial scale ($x_L \sim k^{-1}$). This means that we must treat the linearly unperturbed quantities as a function of $\boldsymbol{\xi}$ and τ. Similarly the complex (Fourier) amplitudes of the excited wave must also be a function of these slowly varying quantities, $A(\xi, \tau)\,\mathrm{e}^{i(\boldsymbol{k}\cdot\boldsymbol{x}-\omega t)}$.

If, however, a wave has an extremely long wavelength λ so that λ becomes comparable to $\xi(\sim (W/n_0 T)^{-1} \lambda')$ where λ' is the wavelength of simultaneously existing waves with shorter wavelength, then the perturbation technique breaks down. One must consider the deformation of the waveform itself from a sinusoidal to a non-sinusoidal form, producing a shock or solitons (Section 4.4). For perturbations of waves with a much shorter wavelength, we can assume that the linear wave number and frequency are not significantly altered, and expressions of the form $A(\xi, \tau)\,\mathrm{e}^{i(\boldsymbol{k}\cdot\boldsymbol{x}-\omega_{\boldsymbol{k}} t)}$ are acceptable, where $\omega_{\boldsymbol{k}} = \omega(\boldsymbol{k})$ is the linear frequency response of the system.

If there are many waves excited in the plasma, we can represent them by the sum of all of these waves;

$$A(\xi, \tau, \boldsymbol{x}, t) = \sum_{\boldsymbol{k}=-\infty}^{\infty} A_{\boldsymbol{k}}(\xi, \tau)\,\mathrm{e}^{i(\boldsymbol{k}\cdot\boldsymbol{x}-\omega_{\boldsymbol{k}} t)} \tag{4.1}$$

where $A_{\boldsymbol{k}}$ is the Fourier amplitude given by

$$\begin{gathered} A_{\boldsymbol{k}}(\xi, \tau)\,\mathrm{e}^{-i\omega_{\boldsymbol{k}} t} = \frac{1}{V}\int d\boldsymbol{x}\, A(\xi, \tau, \boldsymbol{x})\,\mathrm{e}^{-i\boldsymbol{k}\cdot\boldsymbol{x}-i\omega_{\boldsymbol{k}} t} \\ k = \mathrm{n}\pi/L, \qquad \mathrm{n} = 0, \pm 1, \pm 2 \ldots, \end{gathered} \tag{4.2}$$

and V is the plasma volume. We note $A_{-\boldsymbol{k}} = A_{\boldsymbol{k}}^*$, while $\omega_{-\boldsymbol{k}} = -\omega_{\boldsymbol{k}}^*$. In the product of the waves, we note the following convolution relation;

$$\frac{1}{V}\int A(\boldsymbol{x})\,B(\boldsymbol{x})\,e^{-i\boldsymbol{k}\cdot\boldsymbol{x}}\,d\boldsymbol{x} = \sum_{\boldsymbol{k}'=-\infty}^{\infty} A_{\boldsymbol{k}'}\,B_{\boldsymbol{k}-\boldsymbol{k}'}\,. \tag{4.3}$$

It is also convenient to note the following relations:

$$\sum_{\boldsymbol{k}=-\infty}^{\infty} |A_{\boldsymbol{k}}|^2 = \frac{1}{V} \int A(\boldsymbol{x})^2 \, d\boldsymbol{x}, \tag{4.4a}$$

or, in a one dimensional case.

$$\sum_{k=-\infty}^{\infty} |A_k|^2 = \frac{1}{2L} \int_{-L}^{L} A(x)^2 \, dx = \overline{A(x)^2}. \tag{4.4b}$$

In the limit of a very large L, the k spacing, π/L, becomes very small and the spectrum may be treated as continuous. It is then convenient to introduce the spectral density $I(\boldsymbol{k})$ (energy density per unit volume in $\boldsymbol{k}$ space) which relates to $\sum_{\boldsymbol{k}=-\infty}^{\infty} |A_{\boldsymbol{k}}|^2$ by

$$\sum_{\boldsymbol{k}=-\infty}^{\infty} |A_{\boldsymbol{k}}|^2 = \int_{-\infty}^{\infty} I(\boldsymbol{k}) \, d\boldsymbol{k} \tag{4.5}$$

where $d\boldsymbol{k}$ is a volume element in $\boldsymbol{k}$ space. $I(\boldsymbol{k})$ is related to the Fourier spectrum of $A(\boldsymbol{x})$ through

$$I(\boldsymbol{k}) = \lim_{V \to \infty} \frac{1}{V} \frac{1}{(2\pi)^3} |A(\boldsymbol{k})|^2 \tag{4.6}$$

where

$$A(\boldsymbol{k}) = \int_{-\infty}^{\infty} d\boldsymbol{x} \, A(\boldsymbol{x}) \, e^{-i(\boldsymbol{k}\cdot\boldsymbol{x})}. \tag{4.7}$$

The actual method of approach to a nonlinear problem very much depends on the nature of the individual problem, hence we do not generalize the approach too much here but leave it to the individual examples.

4.2 Nonlinear Effects on Particles

4.2a Introduction

When electromagnetic (or electrostatic) waves are excited, particle motions in a plasma are affected by these waves. The nature of the effects of these waves, however, depends on whether the waves can be regarded as "random" or not. The effects of the excited waves can be considered stochastic, if the excited waves have an incoherent spectrum with a spectral width $\Delta\omega$ such that the correlation time $\Delta\omega^{-1}$ is much shorter than the time scale t_{NL} of the considered nonlinear process, i.e., $\Delta\omega^{-1} \ll t_{NL}$. (Of course $\Delta\omega^{-1}$ should be much longer than $\langle\omega\rangle^{-1}$, where $\langle\omega\rangle$ is the average frequency of the excited waves.) This stochastic situation leads to a diffusion of unperturbed as well as perturbed velocity distribution functions both in velocity and coordinate space. We discuss these problems in Subsections 4.2b and 4.2c.

When the excited wave has a narrow spectrum and the correlation time is longer than the nonlinear time scale i.e., $\Delta\omega^{-1} \gg t_{NL}$, the wave remains coherent and one must consider coherent interactions of the wave and particles. This leads, for example, to trapping of particles in the wave potential. Such a problem will be treated in Subsection 4.2d.

4.2b Quasilinear Theory

We consider here effects of instability-excited waves on the unperturbed portion of the particle velocity distribution function. We have already discussed one such effect in Subsection 2.3d. Here we discuss the one dimensional quasilinear diffusion of a distribution function associated with an instability in a two humped velocity distribution (Romanov and Filippov, 1961; Vedenov *et al.*, 1961; Drummond and Pines, 1962). The purpose is to study the way in which such diffusion can lead to the stabilization of the instability, and to compare it with other processes.

We consider an electron distribution function with a hump at $v = v_0$ as shown in Fig. 34. We consider the case where the humped portion has a large spread in velocity, so that the "thermal" velocity of the hump is larger than v_0. (We discuss the alternative case in Subsection 4.2d.) As we have seen, the instability in such a case is generated by the positive gradient of the distribution function ($\partial f_0/\partial v > 0$), rather than by the negative energy wave of the stream. We assume that the instability is one dimensional in that all the excited waves have their wave vectors $\boldsymbol{k}$ directed in the direction of $\boldsymbol{v}_0$ (this is a rather crucial assumption and the result we will obtain here cannot be extended to the case with a three dimensional $\boldsymbol{k}$ spectrum). We further assume that only electron dynamics are involved in the process. This assumption is justified for the study of the linear process because the ion contribution to the linear dispersion relation is negligible due to the small mass ratio, but it presents a severe limitation in the nonlinear process because the excited waves can couple with the ion waves through nonlinear wave-wave or wave-particle interactions. Quasilinear theory ignores such nonlinear wave-wave interaction processes, which is why the name "quasi-linear" is used.

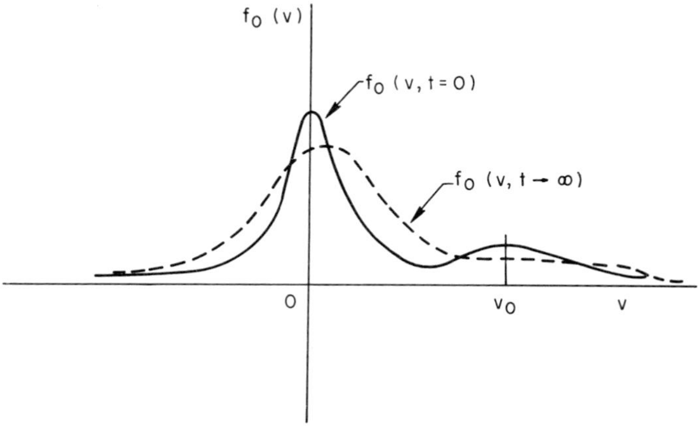

Fig. 34. "Bump in tail" distribution (solid line) and the stabilization by quasilinear diffusion (dotted line)

Under these assumptions, the Vlasov-Maxwell equations for electrons can be simplified to

$$\frac{\partial f}{\partial t}+v\frac{\partial f}{\partial x}-\frac{eE}{m_e}\frac{\partial f}{\partial v}=0, \tag{4.8}$$

$$\frac{dE}{dx}=\frac{en_0}{\varepsilon_0}\left(1-\int_{-\infty}^{\infty} f\,dv\right). \tag{4.9}$$

We write the distribution function f in terms of two portions f_0 and f_1,

$$f=f_0+f_1 \tag{4.10}$$

where f_0 is the spatially averaged distribution function $\langle f\rangle$ which is slowly varying in time, while the perturbed distribution function f_1 represents a set of oscillations with randomly distributed phase. We represent f_1 by (cf.: Eq. (4.1)),

$$f_1(x,v,t)=\sum_{k\neq 0} f_k(v)\,e^{i(kx-\omega_k t)}. \tag{4.11}$$

We use the Fourier series expansion rather than the Fourier transformation simply because f_k has the same dimension as f. This method is convenient for discussing wave-wave interactions. However, for a wave-particle interaction, because the velocity distribution is assumed to be a continuous function of velocity, it is convenient to assume that the k spacing between two adjacent harmonics, $2\pi/L$, is small enough that the summation in Eq. (4.11) may be expressed in terms of an integral. Such a transformation of the expression will be made later. We further assume that the perturbed distribution function f_1 is smaller than f_0 by a factor of ε. We also expand the electric field E in a Fourier series as shown in Eq. (4.11); we assume $E_0=0$ and E_1 to be the same size as f_1.

As was discussed in Subsection 4.1b, nonlinear interactions slowly change these Fourier amplitudes. However, in the quasilinear theory, we assume such a change is negligible compared with the time scale ω_k^{-1}, or exactly speaking, with γ_k^{-1}, where γ_k is the linear growth rate. However, we treat f_0 as a slowly varying function of time with a time scale τ given by

$$\tau=\varepsilon^2 t. \tag{4.12}$$

From the order ε of the Vlasov equation (4.8), we obtain

$$f_k=\frac{e}{m_e}\frac{E_k\,\partial f_0/\partial v}{i(kv-\omega_k)}. \tag{4.13}$$

If we substitute this into the Maxwell equation (4.9), we obtain the following dispersion relation:

$$1-\frac{\omega_p^2}{k^2}\int\frac{\partial f_0/\partial v}{v-\omega_k/k}\,dv=0. \tag{4.14}$$

From the causality requirement, we assume ω_k has a small positive imaginary part as before. If the main part of the distribution function has a narrow velocity spread, Eq. (4.14) can be solved for the real and imaginary parts of ω,

$$\omega_k=\omega_{pe}+i\frac{\pi\omega_{pe}^3}{2k}\int\frac{\partial f_0}{\partial v}\,\delta(kv-\omega_{pe})\,dv\equiv\omega_{pe}+i\gamma_k(\tau). \tag{4.15}$$

Here, because f_0 is a function of τ, γ_k must be regarded as also a function of τ. Let us now consider the order ε^2. The equation for the spatially averaged distribution function f_0 is then

$$\frac{\partial f_0}{\partial\tau}=\frac{e}{m_e}\operatorname{Re}\sum_k E_k\frac{\partial f_k^*}{\partial v}=\operatorname{Re}\left[i\left(\frac{e}{m_e}\right)^2\sum_k|E_k|^2\,e^{2\gamma_k t}\frac{\partial}{\partial v}\cdot\frac{\partial f_0/\partial v}{kv-\omega_k^*}\right]\equiv\frac{\partial}{\partial v}\left(D_v\frac{\partial f_0}{\partial v}\right) \tag{4.16}$$

where the diffusion constant D_v is given by

$$D_v = \sum_k \pi \left(\frac{e}{m_e}\right)^2 |E_k|^2 \, e^{2\gamma_k t} \, \delta(kv - \omega_{pe}). \tag{4.17a}$$

The quasi-linear diffusion equation (4.16) shows a diffusion of the distribution function in the velocity range with a positive gradient, where E_k is excited and grows with a growth rate given by γ_k. Because of the delta function dependency of D in Eq. (4.16), it is more convenient to introduce the spectral density function $I(k, t)$ which gives (cf. Eq. (4.5)),

$$\sum_k |E_k|^2 \, e^{2\gamma_k t} = \int_{-\infty}^{\infty} I(k, t) \, dk. \tag{4.18}$$

The diffusion coefficient in Eq. (4.17 a) is then expressed as

$$D_v = \pi \left(\frac{e}{m_e}\right)^2 \int_{-\infty}^{\infty} I(k) \, \delta(kv - \omega_{pe}) \, dk. \tag{4.17b}$$

Now, from Eq. (4.18),

$$\frac{\partial I(k)}{\partial t} = 2\gamma_k(\tau) \, I(k). \tag{4.19}$$

If we use γ_k given by Eq. (4.15), we have

$$\frac{\partial I(k)}{\partial t} = \frac{\pi \omega_{pe}^3}{k} I(k) \int \frac{\partial f_0}{\partial v} \, \delta(kv - \omega_{pe}) \, dv. \tag{4.20}$$

The set of time differential Eqs. (4.20) and (4.16) with the diffusion coefficient given by Eq. (4.17 b) are the so called quasilinear equations. They are called quasilinear because, even though the change of f_0 depends on the nonlinear product of E and $\partial f/\partial v$, the development of this product $\sim I(k)$ is assumed to be linear as shown in Eq. (4.19). If, however, we consider nonlinear mode couplings (cf. Section 4.3), the time dependency of I has a term proportional to I^2. Neglecting this effect can be justified when $\gamma_k \gg t_{NL}^{-1} \sim (W/n_0 T)\,\omega_k$ as discussed before. The set of quasilinear equations shows that, at an initial stage of the instability, γ_k is given by the positive gradient nature of f_0 due to the hump shown in Fig. 34. This instability increases the spectrum intensity $I(k)$ exponentially. The increased $I(k)$ in turn diffuses the distribution function to produce a lesser value of $\partial f_0/\partial v$, reducing the growth rate γ_k. The rate of increase of $I(k)$ is thus adiabatically reduced and may lead to eventual saturation when $\partial f_0/\partial v=0$, for $v=\omega_{pe}/k$ throughout the range of the phase velocity. This is called a plateau formation of the distribution function, and is characteristic of one dimensional quasilinear diffusion. In three dimensions as we have seen in Subsection 2.3d, there is no plateau formation.

Let us study this process a little further. From Eq. (4.17b) we see

$$D_v \frac{\partial f_0}{\partial v} = \pi \left(\frac{e}{m_e}\right)^2 \int_{-\infty}^{\infty} I(k)\,\delta(kv-\omega_{pe})\,dk \frac{\partial f_0}{\partial v}.$$

If we interchange the integral over k with that over v and use Eq. (4.20),

$$D_v \frac{\partial f_0}{\partial v} = -\left(\frac{e}{m_e}\right)^2 \frac{1}{\omega_{pe} v^3} \frac{\partial I(\omega_{pe}/v)}{\partial t}. \tag{4.21}$$

Substituting Eq. (4.21) into (4.16), we have

$$\frac{\partial f_0(v,\tau)}{\partial \tau} + \frac{\partial}{\partial t}\left\{\frac{\partial}{\partial v}\left[\left(\frac{e}{m_e}\right)^2 \frac{1}{\omega_{pe} v^3} I\left(\frac{\omega_{pe}}{v}\right)\right]\right\} = 0. \tag{4.22a}$$

In other words, there exists a conservation relation between the particle distribution function and the spectral density generated by the instability. The different time scale arises due to the quasilinear (or adiabatic) nature of the problem. If we use this conservation relation, we can estimate the total spectral density generated by the instability. From Eq. (4.22a) we see

$$\int_{v_1}^{v} f_0(v)\,dv + \left(\frac{e}{m_e}\right)^2 \frac{1}{\omega_{pe} v^3} I\left(\frac{\omega_{pe}}{v}\right)\Bigg|_{v_1}^{v} = \text{const. in time.} \tag{4.22b}$$

In the framework of the quasilinear theory $I(\omega_{pe}/v)$ exists only for the portion of velocity which corresponds to $\partial f_0/\partial v>0$. Hence we take v_1 to be the velocity at which $\partial f_0/\partial v=0$ near the hump, where $I(\omega_{pe}/v_1)=0$.

Because, at $t=0$, $I=0$,

$$\begin{aligned}\left(\frac{e}{m_e}\right)^2 \frac{1}{\omega_{pe} v^3} I\left(\frac{\omega_{pe}}{v}\right)(t\to\infty) &= \int_{v_1}^{v} [f_0(t=0) - f_0(t\to\infty)]\,dv \\ &= \text{fractional area of the hump in distribution.}\end{aligned} \tag{4.23}$$

The turbulent level $W/n_0 T$ at the quasilinear saturation can be estimated from this expression. If we write the fractional area of the beam hump by

η (a dimensionless constant $\sim$ beam density/plasma density), Eq. (4.23) gives

$$\eta = \left(\frac{e}{m_e}\right)^2 \frac{1}{\omega_{pe} v_0^3} I\left(\frac{\omega_{pe}}{v_0}\right). \tag{4.24}$$

If we assume the spectral width of the turbulence to be Δk

$$W = \frac{1}{2}\varepsilon_0 \sum_k |E_k|^2 \approx \frac{1}{2}\varepsilon_0 I(k)\Delta k. \tag{4.25}$$

Hence

$$\frac{W}{n_0 T} = \frac{\eta}{2\omega_{pe}} \frac{v_0^3}{v_T^2} \Delta k. \tag{4.26a}$$

Because $v_0 \sim v_T$ and $\Delta k \sim \omega_{pe}/v_0$,

$$\frac{W}{n_0 T} \sim \eta. \tag{4.26b}$$

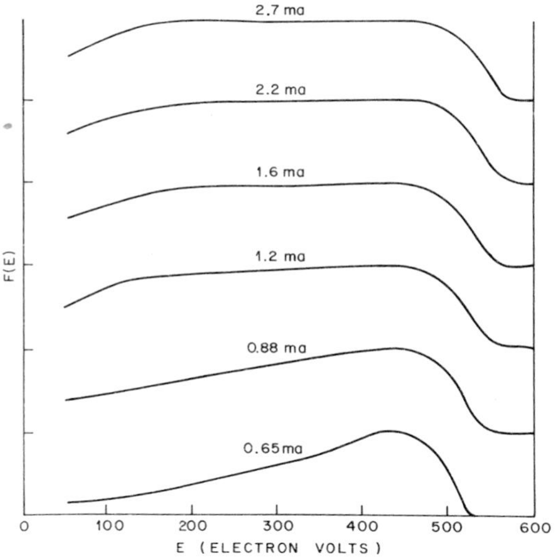

Fig. 35. Distribution function of a weak, broad beam after passing through a plasma. At the lowest current, the instability has not had enough growth time to cause significant diffusion. For the next current, diffusion has begun, but saturation of the energy does not occur within the length of the machine. At higher currents, saturation does occur and the distribution is essentially flat as predicted. (Results only significant for energies above 150 volts. There are unmeasured contributions to the distribution function below that.) (After Roberson and Gentle, 1971)

Thus, the saturation level is given roughly by the ratio of beam density to the plasma density.

Plateau formation in the distribution function as a result of an instability generated by a "bump in the tail" distribution, has been observed experimentally by Roberson and Gentle (1971). The energy distribution function of the bump region is measured at a fixed point as a function of the current carried by the beam as shown in Fig. 35. When the beam current is weak, the instability does not develop to a high intensity, shown by the bottom curve. However, as the beam current is increased, the instability grows faster and the beam distribution is observed to be flattened.

4.2c Resonance Broadening

Let us now consider the effect of fluctuating fields on the perturbed distribution function. As we have seen, for example in Eq. (4.13), the linearized perturbed distribution function $f_{\boldsymbol{k}}$ is singular for velocity $\boldsymbol{v}$, given by $\boldsymbol{k} \cdot \boldsymbol{v} = \omega_{\boldsymbol{k}}$. Hence, ordinary perturbation theory expanded in the power of $\boldsymbol{E}_{\boldsymbol{k}}$ breaks down at this resonant velocity. Consequently, a careful treatment is required to calculate the perturbed distribution function near $\omega = \boldsymbol{k} \cdot \boldsymbol{v}$. For example, effects of particle collisions are important even if the collision frequency ν is much smaller than the wave frequency for the resonant particles for which $|\omega_{\boldsymbol{k}} - \boldsymbol{k} \cdot \boldsymbol{v}| < \nu$. This is because the delta function dependence of $\mathrm{Im}\, f_{\boldsymbol{k}}$, i.e., $\delta(\boldsymbol{k} \cdot \boldsymbol{v} - \omega_{\boldsymbol{k}})$, is no longer valid and may be changed to $(\nu/\pi)[(\omega_{\boldsymbol{k}} - \boldsymbol{k} \cdot \boldsymbol{v})^2 + \nu^2]^{-1}$. This effect is called resonance broadening.

In particular, in the presence of superthermal fluctuations, such resonance broadening is considered to be significant. In that case, the resonance broadening can also be considered to be a consequence of deviation from the unperturbed orbit $x = vt$ to a perturbed orbit due to the fluctuating field. Consequently, the effect is also called the orbit diffusion effect.*

Dupree (1966) was the first to point out the importance of this effect. Dupree's approach has been elaborated by Weinstock (1969). More recently Rudakov and Tsytovich (1971), Kono and Ichikawa (1973), and Mima (1973) have treated the problem using the perturbation theory with a renormalized propagator. As will be seen, an exact treatment of this problem is rather complex, so we show here only qualitatively how such an effect appears and can be evaluated.

---

* Orbit diffusion works also on non-resonant particles. Hence, the resonance broadening is, more precisely, one aspect of the orbit diffusion.

We consider a situation in which a wave is propagating in a medium where there also exist fields with superthermal fluctuations. Such a wave is called a test wave. We assume that the test wave has a wave number $\boldsymbol{k}_0$ (real) and a frequency (complex) ω_0. Then the total oscillating electric field $\boldsymbol{E}(\boldsymbol{x}, t)$ can be expressed by

$$\boldsymbol{E}(\boldsymbol{x}, t)=\boldsymbol{E}^{(t)}(\boldsymbol{x}, t)+\boldsymbol{E}^{(f)}(\boldsymbol{x}, t) \tag{4.27}$$

where $\boldsymbol{E}^{(t)}$ is the electric field of the test wave given by

$$\boldsymbol{E}^{(t)}(\boldsymbol{x}, t)=\boldsymbol{E}_{\boldsymbol{k}_0} \mathrm{e}^{i(\boldsymbol{k}_0 \cdot \boldsymbol{x}-\omega_0 t)}+\text{c.c.} \tag{4.28 a}$$

where c.c. means the complex conjugate, and $\boldsymbol{E}^{(f)}$ is the background fluctuating field given by (cf. Eq. (4.1))

$$\boldsymbol{E}^{(f)}(\boldsymbol{x}, t)=\sum_{\boldsymbol{k}} \boldsymbol{E}_{\boldsymbol{k}} \mathrm{e}^{i(\boldsymbol{k} \cdot \boldsymbol{x}-\omega_{\boldsymbol{k}} t)}. \tag{4.29 a}$$

The particle distribution function $f(\boldsymbol{x}, \boldsymbol{v}, t)$ can be expressed in a similar way;

$$f(\boldsymbol{x}, \boldsymbol{v}, t)=f_0(\boldsymbol{v})+f^{(t)}(\boldsymbol{x}, \boldsymbol{v}, t)+f^{(f)}(\boldsymbol{x}, \boldsymbol{v}, t) \tag{4.30}$$

and

$$f^{(t)}(\boldsymbol{x}, \boldsymbol{v}, t)=f_{\boldsymbol{k}_0} \mathrm{e}^{i(\boldsymbol{k}_0 \cdot \boldsymbol{x}-\omega_0 t)}+\text{c.c.}, \tag{4.28 b}$$

$$f^{(f)}(\boldsymbol{x}, \boldsymbol{v}, t)=\sum_{\boldsymbol{k}} f_{\boldsymbol{k}} \mathrm{e}^{i(\boldsymbol{k} \cdot \boldsymbol{x}-\omega_{\boldsymbol{k}} t)}. \tag{4.29 b}$$

If we substitute these electric fields and distribution functions into the Vlasov equation, we have, for the test wave:

$$i(\boldsymbol{k}_0 \cdot \boldsymbol{v}-\omega_0) f_{\boldsymbol{k}_0}+\frac{q}{m} \boldsymbol{E}_{\boldsymbol{k}_0} \cdot \frac{\partial f_0}{\partial \boldsymbol{v}}+\frac{q}{m} \sum_{\boldsymbol{k}} \boldsymbol{E}_{\boldsymbol{k}_0-\boldsymbol{k}} \cdot \frac{\partial f_{\boldsymbol{k}}}{\partial \boldsymbol{v}} \mathrm{e}^{i \Delta \omega_{0, \boldsymbol{k}} t}=0 \tag{4.31}$$

where $\Delta \omega_{0, \boldsymbol{k}}=\omega_0-(\omega_{\boldsymbol{k}}+\omega_{\boldsymbol{k}_0-\boldsymbol{k}})$ and, for the background fluctuations,

$$i(\boldsymbol{k} \cdot \boldsymbol{v}-\omega_{\boldsymbol{k}}) f_{\boldsymbol{k}}+\frac{q}{m} \boldsymbol{E}_{\boldsymbol{k}} \cdot \frac{\partial f_0}{\partial \boldsymbol{v}}+\frac{q}{m} \sum_{\boldsymbol{k}'} \boldsymbol{E}_{\boldsymbol{k}-\boldsymbol{k}'} \cdot \frac{\partial f_{\boldsymbol{k}'}}{\partial \boldsymbol{v}} \mathrm{e}^{i \Delta \omega_{\boldsymbol{k}, \boldsymbol{k}'} t}=0 \tag{4.32}$$

where $\Delta \omega_{\boldsymbol{k}, \boldsymbol{k}'}=\omega_{\boldsymbol{k}}-(\omega_{\boldsymbol{k}'}+\omega_{\boldsymbol{k}-\boldsymbol{k}'})$. In this set of equations, the nonlinear term (the third term in Eq. (4.31)), which contributes to the test wave through the mode couplings of background fluctuations, contains a term which is proportional to the test wave distribution function $f_{\boldsymbol{k}_0}$ itself through the nonlinear term in Eq. (4.32). In the same way, the nonlinear term in Eq. (4.32) contains a term proportional to $f_{\boldsymbol{k}}$. These terms when combined with the $i(\boldsymbol{k}\cdot\boldsymbol{v}-\omega_{\boldsymbol{k}})\,f_{\boldsymbol{k}}$ term can be seen to modify the resonant effect, $\boldsymbol{k}\cdot\boldsymbol{v}\sim\omega_{\boldsymbol{k}}$. This modification, which is the resonance broadening, should apply to the test wave as well as to the background fluctuations. However, instead of considering the chain of this resonance broadening, we assume here for simplicity that the background fluctuation has a spectrum which is peaked at a wave number $\boldsymbol{k}$ well away from the resonant wave number $\boldsymbol{k}\cdot\boldsymbol{v}_T=\omega_{\boldsymbol{k}}$, where $\boldsymbol{v}_T$ is the thermal velocity. Then we may neglect the resonance broadening effect for the background waves and solve Eq. (4.32) for $f_{\boldsymbol{k}}$ to give:

$$\begin{aligned} f_{\boldsymbol{k}} &= -\frac{q}{m}\,\frac{1}{i(\boldsymbol{k}\cdot\boldsymbol{v}-\omega_{\boldsymbol{k}})}\left(\boldsymbol{E}_{\boldsymbol{k}}\cdot\frac{\partial f_0}{\partial\boldsymbol{v}}+\sum_{\boldsymbol{k}'}\boldsymbol{E}_{\boldsymbol{k}-\boldsymbol{k}'}\cdot\frac{\partial f_{\boldsymbol{k}'}}{\partial\boldsymbol{v}}\,\mathrm{e}^{i\Delta\omega_{\boldsymbol{k},\boldsymbol{k}'}t}\right)\\ &= -\frac{q}{m}\,\frac{1}{i(\boldsymbol{k}\cdot\boldsymbol{v}-\omega_{\boldsymbol{k}})}\left(\boldsymbol{E}_{\boldsymbol{k}}\cdot\frac{\partial f_0}{\partial\boldsymbol{v}}+\boldsymbol{E}_{\boldsymbol{k}-\boldsymbol{k}_0}\cdot\frac{\partial f_{\boldsymbol{k}_0}}{\partial\boldsymbol{v}}\,\mathrm{e}^{i\Delta\omega_{\boldsymbol{k},\boldsymbol{k}_0}t}\right.\\ &\qquad\left.+\sum_{\boldsymbol{k}'\neq\boldsymbol{k}_0}\boldsymbol{E}_{\boldsymbol{k}-\boldsymbol{k}'}\cdot\frac{\partial f_{\boldsymbol{k}'}}{\partial\boldsymbol{v}}\,\mathrm{e}^{i\Delta\omega_{\boldsymbol{k},\boldsymbol{k}'}t}\right) \end{aligned}\tag{4.33}$$

where we have separated the term proportional to $f_{\boldsymbol{k}_0}$.

If we substitute Eq. (4.33) into (4.31), we see that the terms which do not involve $f_{\boldsymbol{k}_0}$ in Eq. (4.33) contribute to modify the linear term $\boldsymbol{E}_{\boldsymbol{k}_0}\cdot\partial f_0/\partial\boldsymbol{v}$ in Eq. (4.31), while the term that contains $f_{\boldsymbol{k}_0}$ modifies $(\boldsymbol{k}_0\cdot\boldsymbol{v}-\omega_0)\,f_{\boldsymbol{k}_0}$. Consequently, we can ignore the former modification compared with the latter. The resultant equation for $f_{\boldsymbol{k}_0}$ then reduces to

$$i(\boldsymbol{k}_0\cdot\boldsymbol{v}-\omega_0)\,f_{\boldsymbol{k}_0}-\frac{\partial}{\partial v_i}\left(D_{ij}\frac{\partial f_{\boldsymbol{k}_0}}{\partial v_j}\right)=-\frac{q}{m}\,\boldsymbol{E}_{\boldsymbol{k}_0}\cdot\frac{\partial f_0}{\partial\boldsymbol{v}},\tag{4.34}$$

where, for electrostatic fluctuations, $\boldsymbol{E}_{\boldsymbol{k}}\|\boldsymbol{k}$, and hence

$$D_{ij}=\mathrm{Re}\left(\frac{q}{m}\right)^2\sum_{\boldsymbol{k}}\frac{k_i\,k_j}{k^2}\,|\boldsymbol{E}_{\boldsymbol{k}}|^2\,\frac{-i}{[(\boldsymbol{k}+\boldsymbol{k}_0)\cdot\boldsymbol{v}-(\omega_{\boldsymbol{k}}+\omega_0)]}.\tag{4.35}$$

We take the real part for D because the imaginary part gives only the frequency shift and does not contribute to the resonance broadening.

We note here that if the resonance broadening were not negligible for the background fluctuations, we would have an expression for $f_{\boldsymbol{k}}$ similar to Eq. (4.34).

We can see now that the nonlinear wave-particle interaction has brought about a diffusion effect in the distribution function for the test wave as shown by the second term in Eq. (4.34).

To solve for the distribution function $f_{\boldsymbol{k}_0}$, we need to know the spectral distribution of the background fluctuation $|\boldsymbol{E}_{\boldsymbol{k}}|^2$. In the case of stabilization of a linear instability, $|\boldsymbol{E}_{\boldsymbol{k}}|^2$ is generated by the linear instability, which is modified by the nonlinear effect considered here. Hence, $|\boldsymbol{E}_{\boldsymbol{k}}|^2$ as a function of $\boldsymbol{k}$ can be found only after solving the complicated integral equation (4.34).

However, to demonstrate the effect of resonance broadening we assume a one dimensional perturbation and postulate a suitable spectrum for the background fluctuation. We postulate that the spectrum is peaked at a much shorter wavelength than that of the test wave, consequently we also assume $\omega_k \gg \omega_0$. Then, by transferring the summation over the Fourier amplitude to the integral over the frequency spectrum density $I(k)$, we have

$$D_{ij} = \pi \left(\frac{q}{m}\right)^2 \int_{-\infty}^{\infty} I(k)\,\delta(kv - \omega_k)\,dk\,, \tag{4.36}$$

where $I(k)$ is given by (cf. Eq. (4.6)),

$$I(k) = \lim_{L\to\infty} \frac{1}{2L}\left(\frac{1}{2\pi}\right)|E(k)|^2 \tag{4.37}$$

and

$$E(k)\,e^{-i\omega_k t} = \int_{-\infty}^{\infty} dx\, E^{(f)}(x)\, e^{-i(kx+\omega_k t)}\,. \tag{4.38}$$

Eq. (4.34) combined with the diffusion constant given by (4.36) formally gives the effect of resonance broadening on the test wave distribution function by the background fluctuations.

Let us consider an example of high frequency ion acoustic turbulence at $\omega \sim \omega_{pi}$ (ion plasma frequency) and $k \sim k_D$ (electron Debye wave number) and its effect on the electron distribution function for a low frequency test wave. In such a case D_{ij} in Eq. (4.36) is given by

$$D_{ij} \equiv D = \pi \left(\frac{e}{m_e}\right)^2 \frac{1}{|v|}\, I\left(\frac{\omega_k}{v}\right). \tag{4.39}$$

By assumption, if the spectrum is peaked at $k \sim k_D$, and $\omega \sim \omega_{pi}$, v may be replaced by its representative value given by

$$v \sim \frac{\omega_{pi}}{k_D} = v_{Te}\left(\frac{m_e}{m_i}\right)^{1/2} \quad (= c_s) \tag{4.40}$$

and the diffusion coefficient can be approximated by

$$D \sim \pi \left(\frac{e}{m_e}\right)^2 \frac{1}{v_{Te}} \left(\frac{m_i}{m_e}\right)^{1/2} I(k_D). \tag{4.41}$$

Because D is the diffusion coefficient in velocity space, it must have the dimension of $[v^2/t]$. For comparison with the classical diffusion coefficient $D^{(c)} \sim v_{Te}^2 \nu_e$, let us evaluate D using the fluctuating field energy density. If we recall that

$$\sum_k E_k^2 = \overline{E(x)^2} = \int_{-\infty}^{\infty} I(k)\, dk, \tag{4.4}$$

and use the fact that $I(k)$ is peaked near $k \sim k_D$,

$$I(k_D) \sim \sum_k E_k^2 / \Delta k \tag{4.42}$$

where Δk is the spectral width. Then the diffusion coefficient can be expressed as

$$D \sim v_{Te}^2 \left(\omega_{pe} \frac{W}{n_0 T}\right) \left(\frac{\pi \omega_{pe}}{\Delta k v_{Te}}\right) \left(\frac{m_i}{m_e}\right)^{1/2} = v_{Te}^2 \nu_{\text{eff}} \tag{4.43}$$

where $W = \varepsilon_0 \sum_k |E_k|^2/2 = \varepsilon_0 \overline{E(x)^2}/2$ is the energy density of the fluctuating field and ν_{eff} is the effective collision frequency in this case and given by (taking $\Delta k \sim k_D$);

$$\nu_{\text{eff}} = \frac{W}{n_0 T} \left(\frac{m_i}{m_e}\right)^{1/2} \cdot \omega_{pe}. \tag{4.44}$$

Because the ion acoustic turbulence has a low phase velocity, it scatters low velocity electrons producing effectively an enhanced anomalous collision effect (cf. Eq. (2.107)). Hence expression (4.43) shows a significantly enhanced diffusion rate.

If we use this diffusion coefficient, we can write the equation for f_{k_0} as

$$i(k_0 v - \omega_0) f_{k_0} - D \frac{\partial^2 f_{k_0}}{\partial v^2} = -\frac{q}{m} E_{k_0} \frac{\partial f_0}{\partial v}. \tag{4.45a}$$

To solve this equation, we apply a Fourier transformation in velocity space:

$$f_{k_0}(v) = \frac{1}{2\pi} \int_{-\infty}^{\infty} f_\tau(\tau)\, e^{i v \tau}\, d\tau, \tag{4.46}$$

where

$$f_\tau(\tau) = \int_{-\infty}^{\infty} f_{k_0}(v)\, e^{-i v \tau}\, dv. \tag{4.47}$$

Then Eq. (4.45a) becomes

$$k_0 \frac{df_\tau}{d\tau} + i\omega_0 f_\tau - \tau^2 D f_\tau = \frac{q}{m} E_{k_0} \overline{\frac{\partial f_0}{\partial v}}, \tag{4.45b}$$

where

$$\overline{\frac{\partial f_0}{\partial v}} = \int_{-\infty}^{\infty} \frac{\partial f_0}{\partial v} e^{-iv\tau} dv. \tag{4.48}$$

Eq. (4.45b) can easily be solved to give

$$f_\tau = \frac{q}{mk_0} E_{k_0} e^{-(i\omega_0\tau - \tau^3 D/3)/k_0} \cdot \int_{-\infty}^{\tau} d\tau' \left[e^{(i\omega_0\tau' - \tau'^3 D/3)/k_0} \cdot \int_{-\infty}^{\infty} dv' \frac{\partial f_0}{\partial v'} e^{-iv'\tau'} \right]. \tag{4.49}$$

Because the diffusion term does not contribute much to the integral for a large τ (due to the exponentially decreasing effect), we approximate $\tau^3 D/3k_0$ by $\tau(D/3k_0)^{1/3}$. Then

$$f_{k_0} = \frac{1}{2\pi} \int f_\tau e^{iv\tau} d\tau \sim -\frac{q}{m} \frac{E_{k_0}}{i(k_0 v - \omega_0) + (k_0^2 D/3)^{1/3}} \frac{\partial f_0}{\partial v}. \tag{4.50}$$

Let us see how this resonance broadening contributes to saturate the ion acoustic wave instability. If we assume cold ions, the resonance broadening appears only for electrons. Recalling the linear instability of an ion acoustic wave by drifting electrons discussed in Subsection 2.1a, the instability is generated by the negative real part of electron conductivity which results from the positive gradient of f_0 at the phase velocity of the wave. If we calculate the real part of the conductivity from the perturbed distribution function given by Eq. (4.50), we see,

$$\mathrm{Re}\,\sigma_e \propto -\int_{-\infty}^{\infty} \frac{\delta}{(\omega_0 - k_0 v)^2 + \delta^2} \frac{\partial f_0}{\partial v} dv \tag{4.51}$$

where $\delta = (k_0^2 D/3)^{1/3}$.

Assuming f_0 to be symmetric about the drift speed v_0, and putting $v - v_0 = u$, we have:

$$\mathrm{Re}\,\sigma_e \propto -\int_{-\infty}^{\infty} \frac{\delta}{[\omega_0 - k_0(v_0 + u)]^2 + \delta^2} \frac{\partial f_0}{\partial u} du. \tag{4.52}$$

Hence $\mathrm{Re}\,\sigma_e$ can be regarded as negligibly small if

$$k_0 v_0 \ll \delta. \tag{4.53a}$$

In view of the fact that for a resonant wave-particle instability (not a negative energy type), $c_s < v_0 < v_{Te}$, we take for a practical saturation limit,

$$\delta = k_0 v_{Te} = \left(\frac{k_0^2 D}{3}\right)^{1/3}. \tag{4.53b}$$

Then, we have

$$\frac{W}{n_0 T} \sim \frac{k_0}{k_D}\sqrt{\frac{m_e}{m_i}} \tag{4.54}$$

which is a fairly low level of saturation.

4.2d Particle Trapping and Wave Overtaking

We now consider a two stream instability that originates from a distribution with a bump in the tail as shown in Fig. 34, but with much narrower spread in distribution for both beam and plasma such that $v_0 \gg v_T$. We still assume the beam density $n_0^{(b)}$ to be much smaller than the background plasma density n_0 and designate the density ratio $n_0^{(b)}/n_0$ by η as before. As was discussed in Subsection 1.3, the instability in such a case is generated by the coupling between the negative energy wave of the beam and the plasma wave in the background plasma. The excited wave has a narrow spectrum and the correlation time $\Delta\omega^{-1}$ is expected to be much longer than the nonlinear time scale t_{NL}, hence we cannot apply the stochastic approach introduced in the previous subsections. However, a semi-analytic approach is possible when the beam to plasma density ration η is regarded as a small parameter (Drummond *et al.*, 1970; Onishchenko *et al.*, 1970; O'Neil *et al.*, 1971; O'Neil and Winfrey, 1972; Gentle and Lohr, 1973). The idea is based on the trapping of the beam electrons in a quasi-monochromatic wave potential excited by the instability. Because of the assumption of the small thermal spread of the distribution function we can use a cold fluid equation for the description of the beam plasma system; referring to Section 1.3, we can write down the dispersion relation for such a case as

$$\frac{\omega_{pe}^2}{\omega^2} + \frac{\omega_{pb}^2}{(\omega - k v_0)^2} = 1, \tag{4.55}$$

where ω_{pe}, ω_{pb} and v_0 are the electron plasma frequency of the plasma, the beam plasma frequency and the drift speed of the beam. The plasma is assumed to be one-dimensional and we have ignored the ion dynamics which may produce limitations in the result of the nonlinear calculation. For a small value of $\eta = \omega_{pb}^2/\omega_{pe}^2$, the temporal growth solution (complex ω)

for a real value of $k=\omega_{pe}/v_0$ is given by

$$\omega=\omega_{pe}(1-\delta) \tag{4.56}$$

where

$$\mathrm{Re}\,\delta=2^{-4/3}\eta^{1/3}\equiv\delta_r, \tag{4.57}$$

$$\mathrm{Im}\,\delta=-3^{1/2}\,2^{-4/3}\eta^{1/3}\equiv\delta_i \tag{4.58}$$

and

$$\delta^3=-\eta/2.$$

The corresponding difference between the beam velocity v_0 and the phase velocity ω_r/k is

$$\Delta v_0=v_0-\frac{\omega_r}{k}=\delta_r v_0>0, \tag{4.59}$$

meaning that the slower wave is the growing wave (cf. Section 1.3). In the linear regime, the phase and the amplitude relations among various dependent variables can be obtained: from the equation of continuity for the beam,

$$\frac{n_1^{(b)}}{n_0^{(b)}}=-\frac{v_1^{(b)}}{\delta_r v_0}; \tag{4.60}$$

from the equation of motion,

$$\frac{v_1^{(b)}}{v_0}=\frac{e}{m_e}\frac{\phi_1}{\delta_r v_0^2}, \tag{4.61}$$

$$\frac{v_1^{(p)}}{v_0}=\frac{e}{m_e}\frac{\phi_1 k}{v_0\omega_{pe}}=-\frac{e}{m_e}\frac{\phi_1}{v_0^2}. \tag{4.62}$$

Superscripts (b) and (p) correspond to the beam and the plasma quantities; v_1, n_1, ϕ_1 are the perturbed fluid velocity, the number density and the field electrostatic potential respectively. Because δ_r is a small quantity, we see from these expressions that $v_1^{(p)}/v_0 \ll v_1^{(b)}/v_0 \ll n_1^{(b)}/n_0^{(b)}$. Specifically the perturbation on the beam density becomes 100% when $|v_1^{(b)}|\sim\delta_r v_0$ and $|v_1^{(p)}|\sim\delta_r^2 v_0$. Note also the phase relation among those variables, in particular the fact that $v_1^{(b)}$ and ϕ_1 are in phase for the slower wave. The idea developed by Drummond *et al.* (1970) and Onishchenko *et al.* (1970) is that because of the narrow spectrum of the excited instability, the wave potential ϕ_1 can effectively be represented by a monochromatic sinusoidal wave $\phi\cos(kx-\omega t)=\phi\cos\omega_{pe}[x/v_0-t(1-\delta_r)]$ until the nonlinearity is developed by the trapping of the beam electrons by this potential.

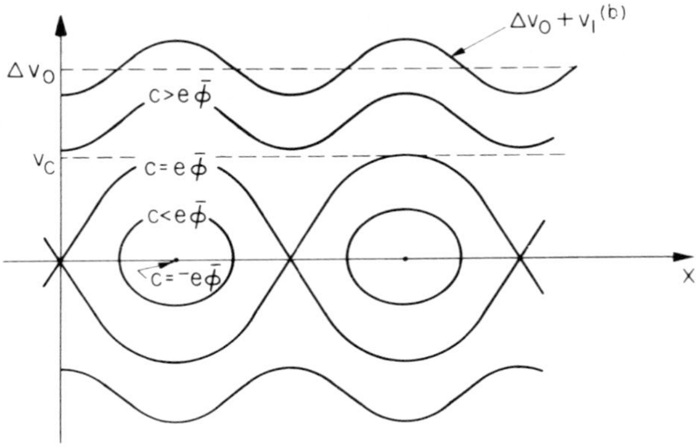

Fig. 36. Electrostatic potential of a wave and its associated velocity contour in phase space

Under this assumption, in the frame moving with the phase velocity of the excited wave $\omega_r/k = \omega_{pe}(1-\delta_r)/k = v_0(1-\delta_r)$, the potential seen by the beam electrons is a stationary sinusoid $\phi_1 = \bar{\phi}\cos kx$. The particle (electron) trajectory in the presence of such a potential is decided by the family of solutions v that satisfy the energy conservation equation

$$\frac{1}{2} m_e v^2 - e\bar{\phi}\cos kx = c, \tag{4.63}$$

where c is the total energy of a particle. Electrons with $c = -e\bar{\phi}$, have $v = 0$, and are trapped at the bottom of the potential well, while those with $c = e\bar{\phi}$ just escape from the well and their phase space trajectory separates the trapped electrons from the untrapped electrons as shown in Fig. 36. The critical velocity of escape v_c in the figure is obtained from $\bar{\phi}$ as

$$v_c^2 = \frac{4e\bar{\phi}}{m_e}. \tag{4.64}$$

Consequently, according to these authors, the trapping of the beam by the potential will start to occur when $\bar{\phi}$ reaches the level given by

$$\bar{\phi}_T = \frac{m_e \Delta v_0^2}{4e} \tag{4.65}$$

and the corresponding field energy density becomes

$$W_1 = 2^{-31/3}\eta^{1/3}(n_0^{(b)} m_e v_0^2/2). \tag{4.66}$$

Even after the trapping, the wave will continue to grow until most of the beam electrons are trapped and rotated in the phase space to the bottom of the potential well. At this time the beam electrons have lost an amount of kinetic energy $\simeq 2 n_0^{(b)} m_e v_0 \Delta v_0$. At this stage, because the background electrons will still have a linear orbit, as can be seen from Eq. (4.62), the wave behavior may still be regarded as linear. Therefore, we can estimate here that half of the lost kinetic energy $\simeq n_0^{(b)} m_e v_0 \Delta v_0$ will go into the field energy and the other half into the kinetic energy of oscillations. The field energy density at this stage therefore can be estimated as being:

$$W_2 \simeq n_0^{(b)} m_e v_0 \Delta v_0 = 2^{-1/3} \eta^{1/3} (n_0^{(b)} m_e v_0^2/2). \tag{4.67}$$

As the trapped particles rotate in the phase space along equipotential contours, they will eventually phase mix due to the difference of the rotating frequency arising from the non-parabolic shape of the potential. After several rotations, the phase space position of the beam particles will smear out, while the field energy will have a damped oscillation with a damping time corresponding to the smearing of the particles in the phase space. The asymptotic value of the field energy density W_f in this state of meta-equilibrium is then given by one-half the difference between the initial beam energy and the beam energy of the smeared out distribution. Since the smeared out distribution is symmetric about the phase velocity of the wave, we find

$$W_f = n_0^{(b)} m_e v_0 \Delta v_0/2 = 2^{-4/3} \eta^{1/3} (n_0^{(b)} m_e v_0^2/2). \tag{4.68}$$

Hence the temporal change of the field energy density $\varepsilon_0 k^2 |\phi|^2/2$ will have an oscillatory form as shown in Fig. 37.

This picture, however, does not present a true process of the dynamics of the trapped beam. First, because of the velocity modulation of the beam, even if v_c reaches the level of Δv_0, full trapping does not occur,

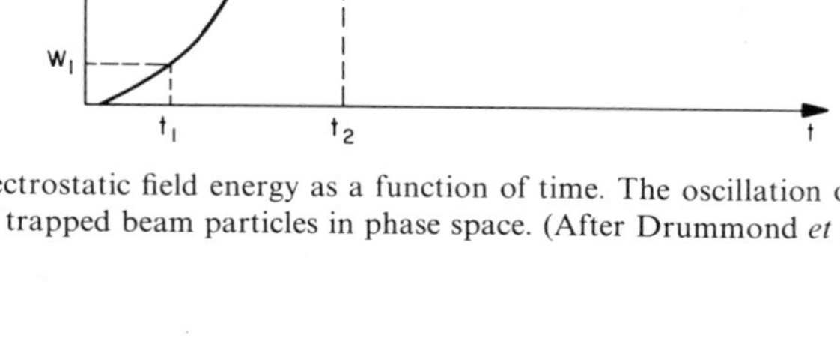

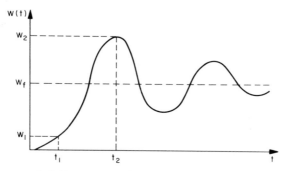

Fig. 37. Electrostatic field energy as a function of time. The oscillation occurs due to the rotation of trapped beam particles in phase space. (After Drummond *et al.*, 1970)

as can be seen in the $v_1^{(b)}$ curve in Fig. 36. Secondly, as will be shown, when trapping occurs, the velocity modulation on the beam $v_1^{(b)}$ reaches Δv_0, hence, according to Eq. (4.60), the beam density modulation $n_1^{(b)}$ reaches $n_0^{(b)}$ i.e., a hundred per cent density modulation. This represents the condition of overtaking (Taniuti, 1963) of the beam and the oscillation pattern of the beam particle should be highly nonlinear.

Let us first prove that $v_1^{(b)} = \Delta v_0$ when the trapping starts. From Fig. 36, we can see that, if we take into account the velocity modulation, the trapping occurs when

$$v_c = \Delta v_0 + v_1^{(b)}, \tag{4.69}$$

namely,

$$\left(\frac{4e\bar{\phi}}{m_e}\right)^{1/2} = \delta_r v_0 + \frac{e}{m_e} \frac{\bar{\phi}}{\delta_r v_0}. \tag{4.70}$$

If we solve this equation for $\bar{\phi}$, we obtain

$$\bar{\phi}'_T = \frac{m_e \Delta v_0^2}{e} \tag{4.71}$$

instead of the expression $\bar{\phi}_T$ in Eq. (4.65). Consequently the amount of velocity modulation corresponding to this value of $\bar{\phi}$ is given from Eq. (4.61) by

$$v_1^{(b)} = \Delta v_0 = \delta_r v_0 \tag{4.72}$$

and the beam density modulation $n_1^{(b)}$ becomes the same as the beam density itself (a 100% modulation).

When such an overtaking occurs, the beam will be bunched together into a small region in phase space. This high density region is trapped in the potential well and will rotate in phase space. The gross picture of the oscillation in wave energy will be the same as shown in Fig. 37, however the details of the phase space dynamics should be different. This fact is seen by O'Neil *et al.* (1971) in their numerical evaluation of the beam dynamics.

Experimental observation of the phase space dynamics of a trapped, weak electron beam by Gentle and Lohr (1973) indicates the bunching of the trapped beam as shown in Fig. 38.

Saturation of the one dimensional ion acoustic wave instability by the trapping of electrons has been also postulated by Nishikawa and Wu (1969). The saturation level becomes extremely small in this case because it occurs when the electron bounce frequency in the excited potential field becomes comparable to the growth rate.

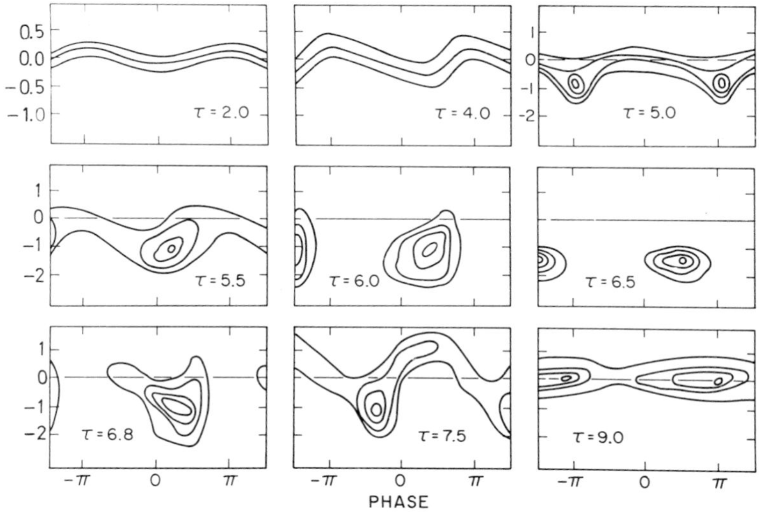

Fig. 38. Experimental contour plots of the electron distribution in phase space. For $\tau = 2, 4$, the central line shows the contour of maximum electron density, with the upper and lower curves indicating half-maximum contours. The width is near the resolution limit of the analyzer. At later τ, the inner contour is the maximum density, normalized to 1.0, and contours are also plotted at 0.8, 0.5, and 0.2. The phase reference is arbitrary and different for each frame. (After Gentle and Lohr, 1973)

4.3 Nonlinear Effects on Waves

4.3a Introduction

In the previous section, we discussed the effect of excited waves on plasma particles. Here we consider how new waves can be excited by nonlinear interactions of the linearly excited waves and the associated saturation mechanism of the instability.

When a wave amplitude grows as a result of a linear instability, its spectrum spreads out because of the nonlinear wave-wave interactions as well as the nonlinear wave-particle interactions. Such diffusion of the wave spectrum tends to limit the growth of the wave amplitude caused by the linear instability. As in the case of effects on particles, the nature of the nonlinear wave-wave interaction differs depending on the coherence of the waves involved.

An interaction is considered "coherent" when the phase of the waves involved in the interaction does not change during the time of the nonlinear interaction t_{NL}. If the spectral width $\Delta\omega$ is known, the condition of a coherent interaction can be approximately written as $t_{NL} < \Delta\omega^{-1}$.

The incoherent case corresponds to the opposite situation where $t_{NL} > \Delta\omega^{-1}$. However, in reality, an apriori judgment of coherence is difficult to make because t_{NL} is not known. Besides, $\Delta\omega^{-1}$ is a dynamic parameter and can change in the process of interactions. In particular, as will be seen, t_{NL} itself depends on the coherence of the process: t_{NL} is short for a coherent interaction $\left(\sim (W/n_0 T)^{-1/2}\omega^{-1}\right)$ and is long for an incoherent interaction $\left(\sim (W/n_0 T)^{-1}\omega^{-1}\right)$; therefore the result does not justify the assumption unless $t_{NL} \gg \Delta\omega^{-1}$ (or $t_{NL} \ll \Delta\omega^{-1}$). The treatment for an intermediate case in the transition from coherence to incoherence is usually very complicated (see for example: Zaslavskii and Sagdeev, 1967).

Here we treat only ideal cases in which the process can be clearly regarded as coherent or incoherent. Subsection 4.3b treats coherent three wave interactions, while Subsection 4.3c treats incoherent wave-wave interactions. On the other hand the wave coherency does not matter for the case of nonlinear wave-particle interactions, in which a wave with wave number $\boldsymbol{k}_1$ and frequency $\omega_{\boldsymbol{k}_1}$ is scattered by a particle with velocity $\boldsymbol{v}$ and transformed into another wave with wave number and frequency, $\boldsymbol{k}_2$ and $\omega_{\boldsymbol{k}_2}$, having a relation $\omega_{\boldsymbol{k}_1} - \omega_{\boldsymbol{k}_2} = (\boldsymbol{k}_1 - \boldsymbol{k}_2) \cdot \boldsymbol{v}$. We discuss such a process in Subsection 4.3d.

4.3b Coherent Three-Wave Interactions

We consider here coherent nonlinear three-wave interactions which are in resonance with each other; that is, for a set of wave numbers $\boldsymbol{k} = \boldsymbol{k}' + \boldsymbol{k}''$, the corresponding frequencies of the three waves satisfy

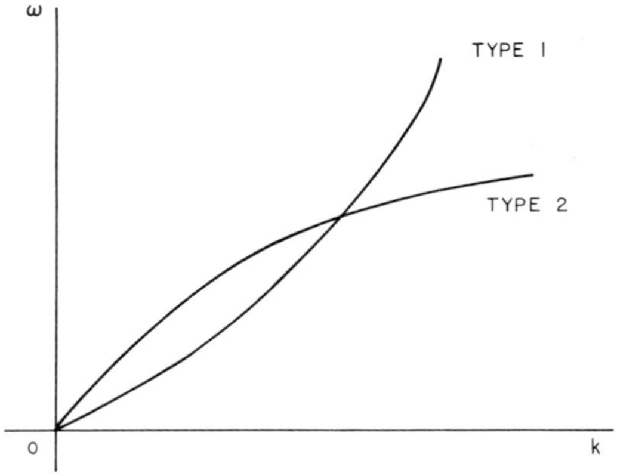

Fig. 39. Dispersion relations for decay type (type 1) and nondecay type (type 2) waves

the frequency matching condition $\omega_{\boldsymbol{k}}=\omega_{\boldsymbol{k}'}+\omega_{\boldsymbol{k}''}$. Such a set of three waves is possible for a wave dispersion curve in $\omega-|k|$ space that is convex downward because $|\boldsymbol{k}'+\boldsymbol{k}''|=|\boldsymbol{k}|\leq|\boldsymbol{k}'|+|\boldsymbol{k}''|$ (type 1 in Fig. 39). A curve that is convex upward as shown by type 2 prohibits such a three wave interaction. As will be seen, when the resonant condition is satisfied, the loss of one quantum of the wave at $\omega_{\boldsymbol{k}}$ produces one quantum at each of $\omega_{\boldsymbol{k}'}$ and $\omega_{\boldsymbol{k}''}$ waves. Hence the type 1 (type 2) dispersion relation is called a decay (nondecay) type.

Many waves in the plasma are of the decay-type. The electron plasma wave is a decay-type when it is combined with the ion acoustic wave. Hence, as a simple example, let us consider electrostatic waves in an unmagnetized hot plasma. We use an electrostatic potential φ in the coupled Vlasov-Poisson equations. We also use Fourier amplitudes $\varphi_{\boldsymbol{k}}(\tau,\xi)\,\mathrm{e}^{-i\omega_{\boldsymbol{k}}t}$ and $f_{\boldsymbol{k}}(\tau,\xi,\boldsymbol{v})\,\mathrm{e}^{-i\omega_{\boldsymbol{k}}t}$ and assume the resonant condition: $\omega_{\boldsymbol{k}}=\omega_{\boldsymbol{k}'}+\omega_{\boldsymbol{k}''}$; then,

$$(\boldsymbol{k}\cdot\boldsymbol{v}-\omega_{\boldsymbol{k}})\,f_{\boldsymbol{k}}^{(j)}=\frac{q_j}{m_j}\,\varphi_{\boldsymbol{k}}\,\boldsymbol{k}\cdot\frac{\partial f_0^{(j)}}{\partial\boldsymbol{v}}+\frac{q_j}{m_j}\sum_{\boldsymbol{k}'\neq\boldsymbol{k}}\varphi_{\boldsymbol{k}'}\,\boldsymbol{k}'\cdot\frac{\partial f_{\boldsymbol{k}-\boldsymbol{k}'}^{(j)}}{\partial\boldsymbol{v}}, \tag{4.73}$$

$$k^2\,\varphi_{\boldsymbol{k}}=\sum_{j=i,e}\frac{q_j n_0}{\varepsilon_0}\int_{-\infty}^{\infty}f_{\boldsymbol{k}}^{(j)}\,d\boldsymbol{v}. \tag{4.74}$$

By iteration of these coupled equations, we can obtain the following nonlinear wave equation for $\varphi_{\boldsymbol{k}}$:

$$\begin{aligned}\varepsilon_{\boldsymbol{k}}^{(1)}\varphi_{\boldsymbol{k}}&+\sum_{\boldsymbol{k}=\boldsymbol{k}'+\boldsymbol{k}''}\varepsilon_{\boldsymbol{k}}^{(2)}(\boldsymbol{k}',\boldsymbol{k}'')\,\varphi_{\boldsymbol{k}'}\varphi_{\boldsymbol{k}''}\\&+\sum_{\boldsymbol{k}=\boldsymbol{k}'+\boldsymbol{k}''+\boldsymbol{k}'''}\varepsilon_{\boldsymbol{k}}^{(3)}(\boldsymbol{k}',\boldsymbol{k}'',\boldsymbol{k}''')\,\varphi_{\boldsymbol{k}'}\varphi_{\boldsymbol{k}''}\varphi_{\boldsymbol{k}'''}+\cdots=0\end{aligned} \tag{4.75}$$

where $\varepsilon^{(1)}$ is the linear dielectric constant given by

$$\varepsilon_{\boldsymbol{k}}^{(1)}=1+\sum_{j=i,e}\frac{\omega_{pj}^2}{k^2}\int d\boldsymbol{v}\,\frac{\boldsymbol{k}\cdot\dfrac{\partial f_0^{(j)}}{\partial\boldsymbol{v}}}{\omega_{\boldsymbol{k}}-\boldsymbol{k}\cdot\boldsymbol{v}+i0}. \tag{4.76}$$

The 2nd and the 3rd order dielectric constants are, after symmetrizing,

$$\begin{aligned}\varepsilon_{\boldsymbol{k}}^{(2)}(\boldsymbol{k}',\boldsymbol{k}'')=&-\frac{1}{2}\sum_{j=i,e}\frac{\omega_{pj}^2}{k^2}\frac{q_j}{m_j}\int d\boldsymbol{v}\,\frac{1}{\omega_{\boldsymbol{k}}-\boldsymbol{k}\cdot\boldsymbol{v}+i0}\\&\cdot\left(\boldsymbol{k}'\cdot\frac{\partial}{\partial\boldsymbol{v}}\,\frac{1}{\omega_{\boldsymbol{k}''}-\boldsymbol{k}''\cdot\boldsymbol{v}+i0}\,\boldsymbol{k}''\cdot\frac{\partial}{\partial\boldsymbol{v}}\right.\\&\left.+\boldsymbol{k}''\cdot\frac{\partial}{\partial\boldsymbol{v}}\,\frac{1}{\omega_{\boldsymbol{k}'}-\boldsymbol{k}'\cdot\boldsymbol{v}+i0}\,\boldsymbol{k}'\cdot\frac{\partial}{\partial\boldsymbol{v}}\right)f_0^{(j)},\end{aligned} \tag{4.77}$$

$$\varepsilon_{\boldsymbol{k}}^{(3)}(\boldsymbol{k}',\boldsymbol{k}'',\boldsymbol{k}''')=\frac{1}{3}\sum_{j=i,e}\frac{\omega_{pj}^2}{k^2}\left(\frac{q_j}{m_j}\right)^2\int d\boldsymbol{v}\frac{1}{\omega_{\boldsymbol{k}}-\boldsymbol{k}\cdot\boldsymbol{v}+i0}$$
$$\cdot\left\{\boldsymbol{k}'\cdot\frac{\partial}{\partial\boldsymbol{v}}\frac{1}{\omega_{\boldsymbol{k}''}+\omega_{\boldsymbol{k}'''}-(\boldsymbol{k}''+\boldsymbol{k}''')\cdot\boldsymbol{v}+i0}\right.$$
$$\cdot\left[\boldsymbol{k}''\cdot\frac{\partial}{\partial\boldsymbol{v}}\frac{1}{\omega_{\boldsymbol{k}'''}-\boldsymbol{k}'''\cdot\boldsymbol{v}+i0}\boldsymbol{k}'''\cdot\frac{\partial}{\partial\boldsymbol{v}}\right. \tag{4.78}$$
$$\left.+\boldsymbol{k}'''\cdot\frac{\partial}{\partial\boldsymbol{v}}\frac{1}{\omega_{\boldsymbol{k}''}-\boldsymbol{k}''\cdot\boldsymbol{v}+i0}\boldsymbol{k}''\cdot\frac{\partial}{\partial\boldsymbol{v}}\right]$$
$$\left.+(2\text{ other permutations of }k\text{'s})\right\}f_0^{(j)}.$$

In these expressions, $\omega_{\boldsymbol{k}}=\omega_{\boldsymbol{k}'}+\omega_{\boldsymbol{k}''}$ and $\omega_{\boldsymbol{k}}=\omega_{\boldsymbol{k}'}+\omega_{\boldsymbol{k}''}+\omega_{\boldsymbol{k}'''}$ are assumed. This causes, unlike previous definition of $\omega_{\boldsymbol{k}}$ (Subsection 4.1 b), $\omega_{\boldsymbol{k}'}$, $\omega_{\boldsymbol{k}''}$ etc. not necessarily to satisfy the linear dispersion relation. Namely for example, $\varepsilon_{\boldsymbol{k}'}^{(1)}(\omega_{\boldsymbol{k}'})$ is not zero, etc.

We now construct coupled three wave equations for an arbitrary pair of waves that satisfy the resonant condition, $\omega_{\boldsymbol{k}_0}=\omega_{\boldsymbol{k}_1}+\omega_{\boldsymbol{k}_2}$. We use the subscripts 0, 1, 2 to designate this specific set of waves and assume $\omega_{\boldsymbol{k}_0}$ is the wave with highest frequency; we take all of $\omega_{\boldsymbol{k}}$'s to be positive. We consider here *resonant three wave interactions*, that is, we assume that these $\omega_{\boldsymbol{k}}$'s satisfy the linear dispersion relation, $\varepsilon_{\boldsymbol{k}}(\omega_{\boldsymbol{k}})=0$. The nonlinear coupling, however, produces small shifts in frequency and wave number from the linear values. These shifts are designated by the τ and ξ dependencies in $\varphi_{\boldsymbol{k}}$. To construct the wave equation for $\varphi_{\boldsymbol{k}}(\tau,\xi)$, we first expand $\varepsilon_{\boldsymbol{k}}^{(1)}$ around $\boldsymbol{k}_0$ and $\mathrm{Re}\,\omega_{\boldsymbol{k}_0}$

$$\varepsilon_{\boldsymbol{k}}^{(1)}=\mathrm{Re}\,\varepsilon_{\boldsymbol{k}_0}^{(1)}+i\,\mathrm{Im}\,\varepsilon_{\boldsymbol{k}_0}^{(1)}+\left.\frac{\partial\varepsilon^{(1)}}{\partial\omega}\right|_{\boldsymbol{k}_0}\Delta\omega+\left.\frac{\partial\varepsilon^{(1)}}{\partial\boldsymbol{k}}\right|_{\boldsymbol{k}_0}\cdot\Delta\boldsymbol{k}$$

and replace $\Delta\omega$ by $i\partial/\partial\tau$ and $\Delta\boldsymbol{k}$ by $-i\partial/\partial\boldsymbol{\xi}$ to give

$$\varepsilon_{\boldsymbol{k}}^{(1)}\varphi_{\boldsymbol{k}_0}=i\frac{\partial\varepsilon}{\partial\omega_0}\left[-\gamma_0+\left(\frac{\partial}{\partial\tau}+\boldsymbol{v}_{g0}\cdot\frac{\partial}{\partial\boldsymbol{\xi}}\right)\right]\varphi_{\boldsymbol{k}_0}(\tau,\xi)$$
$$\equiv i\frac{\partial\varepsilon}{\partial\omega_0}\left(-\gamma_0+\frac{\partial}{\partial\tau'}\right)\varphi_{\boldsymbol{k}_0}(\tau,\xi) \tag{4.79}$$

where

$$\boldsymbol{v}_{g0}=\partial\omega_{\boldsymbol{k}}/\partial\boldsymbol{k}|_{\boldsymbol{k}=\boldsymbol{k}_0}, \tag{4.80}$$

$$\frac{\partial \varepsilon}{\partial \omega_0} = \frac{\partial \operatorname{Re} \varepsilon_{\boldsymbol{k}}^{(1)}}{\partial \omega_{\boldsymbol{k}}}\bigg|_{\boldsymbol{k}=\boldsymbol{k}_0}, \tag{4.81}$$

$$\gamma_0 = -\operatorname{Im} \varepsilon_{\boldsymbol{k}_0}^{(1)}/(\partial \varepsilon/\partial \omega_0). \tag{4.82}$$

Note here that $\varepsilon_{\boldsymbol{k}}^{(1)}$ contains the dielectric constant of free space; $\varepsilon_{\boldsymbol{k}}^{(1)} = 1 + \varepsilon$, and

$$\frac{\partial}{\partial \tau'} = \frac{\partial}{\partial \tau} + \boldsymbol{v}_{g0} \cdot \frac{\partial}{\partial \xi}. \tag{4.83}$$

In these expressions, $\boldsymbol{v}_{g_0}$ and γ_0 are the group velocity and the linear growth rate of the $\omega_{\boldsymbol{k}_0}$ wave respectively. If we substitute Eq. (4.79) into Eq. (4.75) and take only to the order $\varepsilon^{(2)}$, we have the following coupled equations;

$$\begin{aligned}
i\left(\frac{\partial}{\partial \tau'} - \gamma_0\right) \varphi_{\boldsymbol{k}_0} \frac{\partial \varepsilon}{\partial \omega_0} &= -\varepsilon_{\boldsymbol{k}_0}^{(2)}(\boldsymbol{k}_1, \boldsymbol{k}_2)\, \varphi_{\boldsymbol{k}_1} \varphi_{\boldsymbol{k}_2} \\
i\left(\frac{\partial}{\partial \tau'} - \gamma_1\right) \varphi_{\boldsymbol{k}_1} \frac{\partial \varepsilon}{\partial \omega_1} &= -\varepsilon_{\boldsymbol{k}_1}^{(2)}(-\boldsymbol{k}_2, \boldsymbol{k}_0)\, \varphi_{-\boldsymbol{k}_2} \varphi_{\boldsymbol{k}_0} \\
i\left(\frac{\partial}{\partial \tau'} - \gamma_2\right) \varphi_{\boldsymbol{k}_2} \frac{\partial \varepsilon}{\partial \omega_2} &= -\varepsilon_{\boldsymbol{k}_2}^{(2)}(\boldsymbol{k}_0, -\boldsymbol{k}_1)\, \varphi_{\boldsymbol{k}_0} \varphi_{-\boldsymbol{k}_1}.
\end{aligned} \tag{4.84}$$

Here we note $\varphi_{-\boldsymbol{k}} = \varphi_{\boldsymbol{k}}^*$. Now we symmetrize these equations by using the normalized complex amplitude defined by

$$A_{\boldsymbol{k}_j}(\tau, \xi) = \varphi_{\boldsymbol{k}_j}(\tau, \xi) \left| \frac{k_j^2}{2\omega_{\boldsymbol{k}_j}} \frac{\partial(\omega\, \varepsilon_0\, \varepsilon)}{\partial \omega_j} \right|^{1/2} \tag{4.85}$$

and use a quantity $S_{\boldsymbol{k}}$ that designates the sign of the energy of the $\boldsymbol{k}$ wave:

$$\begin{aligned}
S_{\boldsymbol{k}_j} &= \operatorname{sign} \frac{\partial \varepsilon}{\partial \omega_j} \\
&= \operatorname{sign} \frac{1}{\omega_{\boldsymbol{k}}} \frac{\partial(\omega\, \varepsilon_{\boldsymbol{k}}^{(1)})}{\partial \omega}\bigg|_{\boldsymbol{k}_j} \\
&\propto \operatorname{sign} \frac{W_{\boldsymbol{k}_j}}{\omega_{\boldsymbol{k}_j}}
\end{aligned} \tag{4.86}$$

where $W_{\boldsymbol{k}}$ is the energy density of the wave with the wave vector $\boldsymbol{k}$. Then

$$|A_{\boldsymbol{k}}|^2\, S_{\boldsymbol{k}} = \frac{W_{\boldsymbol{k}}}{\omega_{\boldsymbol{k}}} = \hbar\, N_{\boldsymbol{k}} \tag{4.87}$$

represents the action of the $\boldsymbol{k}$ wave and $N_{\boldsymbol{k}}$ is the number density of quanta of the wave. Then the Eq. (4.84) can be written as

$$
\begin{aligned}
& i S_{\boldsymbol{k}_0}\left(\frac{\partial}{\partial \tau'}-\gamma_0\right) A_{\boldsymbol{k}_0}=V_{\boldsymbol{k}_0}(\boldsymbol{k}_1, \boldsymbol{k}_2) A_{\boldsymbol{k}_1} A_{\boldsymbol{k}_2} \\
& i S_{\boldsymbol{k}_1}\left(\frac{\partial}{\partial \tau'}-\gamma_1\right) A_{\boldsymbol{k}_1}=V_{\boldsymbol{k}_1}(-\boldsymbol{k}_2, \boldsymbol{k}_0) A_{-\boldsymbol{k}_2} A_{\boldsymbol{k}_0} \\
& i S_{\boldsymbol{k}_2}\left(\frac{\partial}{\partial \tau'}-\gamma_2\right) A_{\boldsymbol{k}_2}=V_{\boldsymbol{k}_2}(\boldsymbol{k}_0, -\boldsymbol{k}_1) A_{\boldsymbol{k}_0} A_{-\boldsymbol{k}_1}
\end{aligned}
\tag{4.88}
$$

where

$$
V_{\boldsymbol{k}}(\boldsymbol{k}', \boldsymbol{k}'')=-\frac{\varepsilon_0 k^2}{2} \frac{\varepsilon_{\boldsymbol{k}}^{(2)}(\boldsymbol{k}', \boldsymbol{k}'')}{\left|\frac{k^2 \varepsilon_0}{2} \left.\frac{\partial \varepsilon}{\partial \omega}\right|_{\boldsymbol{k}} \frac{(k')^2 \varepsilon_0}{2} \left.\frac{\partial \varepsilon}{\partial \omega}\right|_{\boldsymbol{k}'} \frac{(k'')^2 \varepsilon_0}{2} \left.\frac{\partial \varepsilon}{\partial \omega}\right|_{\boldsymbol{k}''}\right|^{1/2}}. \tag{4.89}
$$

The coupling coefficients $V_{\boldsymbol{k}}(\boldsymbol{k}', \boldsymbol{k}'')$ expressed in this manner have the following symmetric properties. First, by inspecting the expression (4.77), we can easily see

$$
V_{\boldsymbol{k}}(\boldsymbol{k}', \boldsymbol{k}'')=V_{\boldsymbol{k}}(\boldsymbol{k}'', \boldsymbol{k}')=V^*_{-\boldsymbol{k}}(-\boldsymbol{k}', -\boldsymbol{k}''). \tag{4.90}
$$

In addition, if we neglect the wave particle interactions and take contributions only from the principal values of the integral for $\varepsilon_{\boldsymbol{k}}^{(2)}$, $\varepsilon_{\boldsymbol{k}}^{(2)}$ becomes real and we can show the following symmetry properties (for example, see Davidson 1972),

$$
V_{\boldsymbol{k}_1}(-\boldsymbol{k}_2, \boldsymbol{k}_0)=V_{\boldsymbol{k}_2}(\boldsymbol{k}_0, -\boldsymbol{k}_1)=V_{\boldsymbol{k}_0}(\boldsymbol{k}_1, \boldsymbol{k}_2). \tag{4.91}
$$

To study the properties of the coupled Eqs. (4.88), we take the lossless case $\operatorname{Im} \varepsilon_{\boldsymbol{k}}^{(1)}=0$ (hence $\gamma_j=0$) and consider only temporal developments of the amplitudes of the three waves (put $\tau' \to \tau$). Then, in view of the symmetry relations (4.90) and (4.91), if we write $V=V_{\boldsymbol{k}_0}(\boldsymbol{k}_1, \boldsymbol{k}_2)$, we have

$$
\begin{aligned}
& i S_{\boldsymbol{k}_0} \frac{\partial A_{\boldsymbol{k}_0}}{\partial \tau}=V A_{\boldsymbol{k}_1} A_{\boldsymbol{k}_2} \\
& i S_{\boldsymbol{k}_1} \frac{\partial A_{\boldsymbol{k}_1}}{\partial \tau}=V A_{\boldsymbol{k}_0} A_{\boldsymbol{k}_2}^* \\
& i S_{\boldsymbol{k}_2} \frac{\partial A_{\boldsymbol{k}_2}}{\partial \tau}=V A_{\boldsymbol{k}_1}^* A_{\boldsymbol{k}_0}.
\end{aligned}
\tag{4.92a}
$$

We can express the complex amplitude in terms of the real quantities $|A_{\boldsymbol{k}}|$ and $\Theta_{\boldsymbol{k}}$ as

$$A_{\boldsymbol{k}}=|A_{\boldsymbol{k}}|(\tau)\, e^{i\Theta_{\boldsymbol{k}}(\tau)}. \tag{4.93}$$

Eqs. (4.92 a) further reduces to

$$\begin{aligned}
S_{\boldsymbol{k}_0}\frac{\partial |A_{\boldsymbol{k}_0}|}{\partial\tau} &= -V|A_{\boldsymbol{k}_1}||A_{\boldsymbol{k}_2}|\sin\Theta\\
S_{\boldsymbol{k}_1}\frac{\partial |A_{\boldsymbol{k}_1}|}{\partial\tau} &= V|A_{\boldsymbol{k}_0}||A_{\boldsymbol{k}_2}|\sin\Theta\\
S_{\boldsymbol{k}_2}\frac{\partial |A_{\boldsymbol{k}_2}|}{\partial\tau} &= V|A_{\boldsymbol{k}_0}||A_{\boldsymbol{k}_1}|\sin\Theta\\
\frac{\partial\Theta}{\partial\tau} &= V\left(\frac{|A_{\boldsymbol{k}_0}||A_{\boldsymbol{k}_2}|}{S_{\boldsymbol{k}_1}|A_{\boldsymbol{k}_1}|}+\frac{|A_{\boldsymbol{k}_0}||A_{\boldsymbol{k}_1}|}{S_{\boldsymbol{k}_2}|A_{\boldsymbol{k}_2}|}-\frac{|A_{\boldsymbol{k}_1}||A_{\boldsymbol{k}_2}|}{S_{\boldsymbol{k}_0}|A_{\boldsymbol{k}_0}|}\right)\cos\Theta\\
&=\cot\Theta\,\frac{\partial}{\partial\tau}\ln(|A_{\boldsymbol{k}_0}||A_{\boldsymbol{k}_1}||A_{\boldsymbol{k}_2}|)
\end{aligned} \tag{4.92 b}$$

where

$$\Theta=\Theta_{\boldsymbol{k}_0}-\Theta_{\boldsymbol{k}_1}-\Theta_{\boldsymbol{k}_2}.$$

From Eqs. (4.92b), we can find the following constants of motions. First from the imaginary part,

$$|A_{\boldsymbol{k}_0}||A_{\boldsymbol{k}_1}||A_{\boldsymbol{k}_2}|\cos\Theta=\text{const}. \tag{4.94}$$

Second, by multiplying each of the first three equations in (4.92b) by $|A_{\boldsymbol{k}}$'s$|$ adding them together and using the frequency matching condition,

$$W_{\boldsymbol{k}_0}+W_{\boldsymbol{k}_1}+W_{\boldsymbol{k}_2}=\text{const}, \tag{4.95}$$

where $W_{\boldsymbol{k}}$ is the energy density given by Eq. (4.87). Finally, by integrating the two sets of equations in (4.92), we have

$$N_{\boldsymbol{k}_0}+N_{\boldsymbol{k}_1}=\text{const}, \tag{4.96}$$

$$N_{\boldsymbol{k}_0}+N_{\boldsymbol{k}_2}=\text{const}, \tag{4.97}$$

$$N_{\boldsymbol{k}_1}-N_{\boldsymbol{k}_2}=\text{const}. \tag{4.98}$$

From Eqs. (4.96) to (4.98), we can see that when one quantum disappears from the $\omega_{\boldsymbol{k}_0}$ mode, one quantum appears in each of the $\omega_{\boldsymbol{k}_1}$ and $\omega_{\boldsymbol{k}_2}$ modes, i.e., $\Delta N_{\boldsymbol{k}_0}=-1$, $\Delta N_{\boldsymbol{k}_1}=\Delta N_{\boldsymbol{k}_2}=1$. This process is similar to a decay

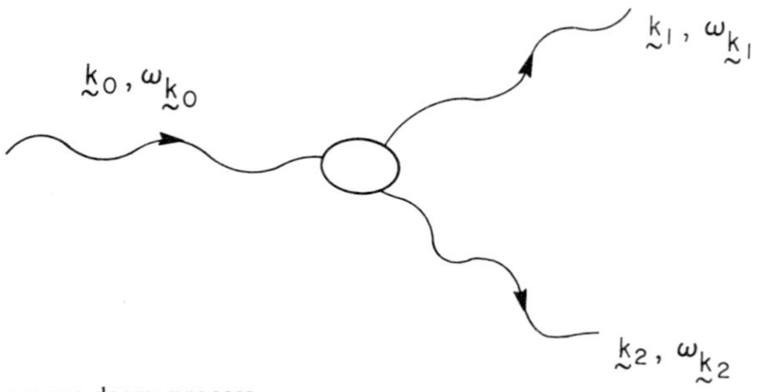

Fig. 40. Three wave decay process

of one particle into two other particles, and it is called the three-wave decay process. This is represented by the diagram in Fig. 40.

Now, let us consider how such a decay process might lead to the stabilization of a linear instability. For the above process to apply, *the excited wave must have a narrow spectrum.* We assume also that the linear instability is not a negative energy type. We take the $\omega_{\boldsymbol{k}_0}$ wave to be the unstable mode. When the unstable wave reaches a certain amplitude, it will excite $\omega_{\boldsymbol{k}_1}$ and $\omega_{\boldsymbol{k}_2}$ waves as shown by Eqs. (4.92b). If we linearize the second and the third equations, assuming $A_{\boldsymbol{k}_0}$ is large and quasistatic, the growth rate of $|A_{\boldsymbol{k}_1}|$ and $|A_{\boldsymbol{k}_2}|$ becomes approximately $V|A_{\boldsymbol{k}_0}|$, i.e., $|A_{\boldsymbol{k}_{1,2}}| \sim \exp(V|A_{\boldsymbol{k}_0}|)\tau$. When we substitute this into the first equation of (4.92b) we can see how such rapidly growing $|A_{\boldsymbol{k}_{1,2}}|$ waves can stop the linear growth of the $A_{\boldsymbol{k}_0}$ mode. This process can also be understood from the energy conservation relation in Eq. (4.95).

The set of equations in (4.92b) admits an exact solution expressed in terms of the elliptic function. For example, let us take as the initial condition,

$$\tau = 0, \qquad |A_{\boldsymbol{k}_0}|^2 \equiv N_0 \gg |A_{\boldsymbol{k}_1}|^2 \equiv N_1$$
$$|A_{\boldsymbol{k}_2}|^2 = 0.$$

Then the dynamic behavior of the $|A_{\boldsymbol{k}}|^2$'s are (see for example, Sagdeev and Galeev, 1969),

$$\begin{aligned}
|A_{\boldsymbol{k}_0}|^2 &= N_0 \,\mathrm{sn}^2 [|V|(\tau-\tau_0)(N_1+N_0)^{1/2}, \eta] \\
|A_{\boldsymbol{k}_1}|^2 &= N_1 + N_0 \{1 - \mathrm{sn}^2 [|V|(\tau-\tau_0)(N_1+N_0)^{1/2}, \eta]\} \\
|A_{\boldsymbol{k}_2}|^2 &= N_0 \{1 - \mathrm{sn}^2 [|V|(\tau-\tau_0)(N_1+N_0)^{1/2}, \eta]\}
\end{aligned} \tag{4.99}$$

where

$$1 - \eta^2 = \frac{N_1}{N_0 + N_1} \ll 1,$$

$x = \mathrm{sn}(u, k)$ is the elliptic function:

$$u = \int_0^x \frac{dx}{\sqrt{(1-x^2)(1-k^2x^2)}},$$

and τ_0 is the time at which $|A_{\boldsymbol{k}_0}|^2$ goes to zero. The temporal behavior of the above solutions is shown in Fig. 41. Because of the periodic nature of the elliptic function, the decay process in this case (i.e., coherent three wave interactions) is *reversible*.

If we also introduce the spatial variation of the actions $|A_{\boldsymbol{k}}|^2$, the left hand side of the coupled Eqs. (4.92) becomes

$$iS_{\boldsymbol{k}_j}\left(\frac{\partial A_{\boldsymbol{k}_j}}{\partial \tau} - \boldsymbol{v}_{g_j}\cdot\frac{\partial A_{\boldsymbol{k}_j}}{\partial \xi}\right); \qquad j = 0, 1, 2.$$

The coupled equations represent propagations of the nonlinearly coupled three-waves. In the case of one dimensional propagation for all the three-waves, Nozaki and Taniuti (1973) have shown that the coupled equations admit solutions similar to those in the self-induced transparency (Lamb, 1971). When a condition is met among the relative intensities of the waves, two of the wave envelopes admit steady state pulse solutions (envelope solitons) and the third a steady state shock (envelope shock) solution; all propagate at a common speed.

Finally if the $\omega_{\boldsymbol{k}_0}$ wave is of the negative energy type (but the rest are positive energy waves), the above situation changes completely because the loss of a quantum from the $\omega_{\boldsymbol{k}_0}$ wave means the gain of a negative quantum to this wave. Consequently (as can also be seen from Eq. (4.95)),

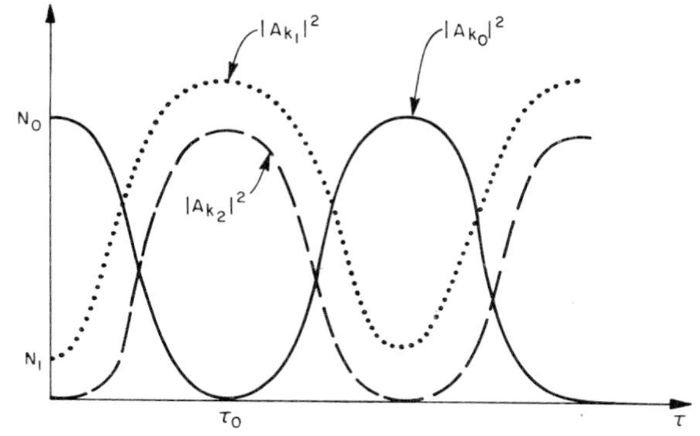

Fig. 41. Temporal behavior of the actions $|A_{\boldsymbol{k}}|^2$ of the three waves for the case when $|A_{\boldsymbol{k}_0}|^2$ has the largest amplitude at $\tau = 0$. (After Sagdeev and Galeev, 1969)

if $W_{k_0} < 0$, all three waves grow simultaneously. The growth rate is larger than exponential. Such a process is called an explosive instability. Thus, the three wave interaction cannot be a saturation mechanism for a linear instability caused by a negative energy wave. We also note that the present theory applies only in the case of *coherent* three-wave interactions; the linearly excited wave must have a narrow spectrum $\Delta\omega$ such that

$$\Delta\omega \ll t_{NL}^{-1} \sim |V|\,|A_{k_0}|.$$

This is because if the phase of A_{k_0} changes often during the time t_{NL}, the average nonlinear growth rate will vanish.

4.3c Three Wave Interactions with Random Phase

We consider here nonlinear three wave interactions in which the phase of the three waves changes rapidly at random times during the process of the interaction. In such a case, the time-averaged amplitude function $\langle A_k \rangle$ vanishes, when averaged over a time T which is much larger than the average period of the phase change but smaller than the nonlinear interaction time scale t_{NL}. This occurs because

$$\langle A_k \rangle = \frac{1}{T} \int_0^T A_k(\tau)\, e^{i\Theta_k(t)}\, dt,$$

where $\Theta_k(t)$ represents the phase of $A_k(\tau)$ which is rapidly varying compared with the time scale of τ, hence

$$\langle A_k \rangle = A_k(\tau) \langle \cos \Theta_k(t) + i \sin \Theta_k(t) \rangle = 0. \tag{4.100}$$

As was discussed, however, the judgment of whether or not the phase changes rapidly during t_{NL} is difficult to make in reality because t_{NL} itself depends on whether the phase changes rapidly during t_{NL}. Let us assume here that such a postulate is justifiable.

In such a case, the use of the complex Fourier amplitude $A_k(\tau)$ does not make sense and only the phase independent quantity $|A_k|^2$ is useful. When the nonlinear interaction starts to take place, however, such a randomly changing phase of the three waves will begin to be correlated. If we write A_k as a sum of the terms with a completely random phase $A_k^{(0)}$ and a small term but with a correlated phase $A_k^{(1)}$,

$$A_k = A_k^{(0)} + A_k^{(1)}$$

where

$$\langle A_{\boldsymbol{k}}^{(0)} \rangle = 0,$$

the average of the product of the three wave amplitudes becomes

$$\begin{aligned}\langle A_{\boldsymbol{k}} A_{\boldsymbol{k}'} A_{\boldsymbol{k}''} \rangle &= \langle A_{\boldsymbol{k}}^{(0)} A_{\boldsymbol{k}'}^{(0)} A_{\boldsymbol{k}''}^{(0)} \rangle + \langle A_{\boldsymbol{k}}^{(1)} A_{\boldsymbol{k}'}^{(0)} A_{\boldsymbol{k}''}^{(0)} \rangle + \langle A_{\boldsymbol{k}}^{(0)} A_{\boldsymbol{k}'}^{(1)} A_{\boldsymbol{k}''}^{(0)} \rangle \\ &\quad + \langle A_{\boldsymbol{k}}^{(0)} A_{\boldsymbol{k}'}^{(0)} A_{\boldsymbol{k}''}^{(1)} \rangle \\ &= \langle A_{\boldsymbol{k}} A_{\boldsymbol{k}'}^{(0)} A_{\boldsymbol{k}''}^{(0)} \rangle + \langle A_{\boldsymbol{k}}^{(0)} A_{\boldsymbol{k}'} A_{\boldsymbol{k}''}^{(0)} \rangle + \langle A_{\boldsymbol{k}}^{(0)} A_{\boldsymbol{k}'}^{(0)} A_{\boldsymbol{k}''} \rangle .\end{aligned} \tag{4.101}$$

The coupled Eqs. (4.88) require modification in this case because of the following two reasons. First, because of a large bandwidth of the spectrum of each of the waves, we cannot select the only three discrete waves but we must sum up over many waves that satisfy the selection rule $\boldsymbol{k} = \boldsymbol{k}' + \boldsymbol{k}''$.

Secondly, in the presence of such a finite bandwidth we must allow a frequency mismatch, i.e.,

$$\omega_{\boldsymbol{k}} - \omega_{\boldsymbol{k}'} - \omega_{\boldsymbol{k}''} \equiv \Delta\omega_{\boldsymbol{k}} \neq 0 .$$

If we consider these points and if, for simplicity, we take only temporal development, Eqs. (4.88) become,

$$i S_{\boldsymbol{k}_0} \frac{\partial A_{\boldsymbol{k}_0}}{\partial \tau} = \sum_{\boldsymbol{k}_0 = \boldsymbol{k}' + \boldsymbol{k}''} V_{\boldsymbol{k}_0}(\boldsymbol{k}', \boldsymbol{k}'') A_{\boldsymbol{k}'} A_{\boldsymbol{k}''} \mathrm{e}^{i\Delta\omega_0 t}, \tag{4.102}$$

$$i S_{\boldsymbol{k}_1} \frac{\partial A_{\boldsymbol{k}_1}}{\partial \tau} = \sum_{\boldsymbol{k}_1 = \boldsymbol{k} - \boldsymbol{k}''} V_{\boldsymbol{k}_1}(-\boldsymbol{k}'', \boldsymbol{k}) A_{-\boldsymbol{k}''} A_{\boldsymbol{k}} \mathrm{e}^{-i\Delta\omega_1 t}, \tag{4.103}$$

$$i S_{\boldsymbol{k}_2} \frac{\partial A_{\boldsymbol{k}_2}}{\partial \tau} = \sum_{\boldsymbol{k}_2 = \boldsymbol{k} - \boldsymbol{k}'} V_{\boldsymbol{k}_2}(\boldsymbol{k}, -\boldsymbol{k}') A_{\boldsymbol{k}} A_{-\boldsymbol{k}'} \mathrm{e}^{-i\Delta\omega_2 t} \tag{4.104}$$

where $\Delta\omega_j$ is the frequency mismatch given by

$$\begin{aligned}\Delta\omega_0 &= \omega_{\boldsymbol{k}_0} - \omega_{\boldsymbol{k}'} - \omega_{\boldsymbol{k}''}, \\ \Delta\omega_1 &= \omega_{\boldsymbol{k}} - \omega_{\boldsymbol{k}_1} - \omega_{\boldsymbol{k}''}, \\ \Delta\omega_2 &= \omega_{\boldsymbol{k}} - \omega_{\boldsymbol{k}'} - \omega_{\boldsymbol{k}_2}.\end{aligned} \tag{4.105}$$

If we recall that the τ dependency of $|A_{\boldsymbol{k}_j}|^2$ originates from the phase correlation arising from the nonlinear interaction, then $A_{\boldsymbol{k}}$ in the lowest order can be obtained by integrating the coupled Eqs. (4.102) to (4.104) from $t = -\infty$ to t, neglecting the τ dependency of the amplitude on the right hand side:

$$A_{\boldsymbol{k}_0} = -S_{\boldsymbol{k}_0} \sum_{\boldsymbol{k}_0=\boldsymbol{k}'+\boldsymbol{k}''} \frac{V_{\boldsymbol{k}_0}(\boldsymbol{k}',\boldsymbol{k}'')A_{\boldsymbol{k}'}^{(0)}A_{\boldsymbol{k}''}^{(0)}}{\Delta\omega_0 - i0} \mathrm{e}^{i\Delta\omega_0 t}, \tag{4.106}$$

$$A_{\boldsymbol{k}_1} = S_{\boldsymbol{k}_1} \sum_{\boldsymbol{k}_1=\boldsymbol{k}-\boldsymbol{k}''} \frac{V_{\boldsymbol{k}_1}(-\boldsymbol{k}'',\boldsymbol{k})A_{\boldsymbol{k}''}^{(0)*}A_{\boldsymbol{k}}^{(0)}\mathrm{e}^{-i\Delta\omega_1 t}}{\Delta\omega_1 + i0}, \tag{4.107}$$

$$A_{\boldsymbol{k}_2} = S_{\boldsymbol{k}_2} \sum_{\boldsymbol{k}_2=\boldsymbol{k}-\boldsymbol{k}'} \frac{V_{\boldsymbol{k}_2}(\boldsymbol{k},-\boldsymbol{k}')A_{\boldsymbol{k}}^{(0)}A_{\boldsymbol{k}'}^{(0)*}\mathrm{e}^{-i\Delta\omega_2 t}}{\Delta\omega_2 + i0}, \tag{4.108}$$

where $\pm i0$ is added to the $\Delta\omega$'s to make the integral convergent at $t=-\infty$. This causality requirement on $\mathrm{Im}\,\Delta\omega$ is very important as will be seen. With these preparations, we now construct the time differential equation for the $|A_{\boldsymbol{k}}|^2$'s. Let us first consider $|A_{\boldsymbol{k}_0}|^2$. If we multiply Eq. (4.102) by $A_{\boldsymbol{k}_0}^*$ and add to it the product of $A_{\boldsymbol{k}_0}$ and the complex conjugate of (4.102), we have

$$\begin{aligned}\frac{\partial|A_{\boldsymbol{k}_0}|^2}{\partial\tau} = -iS_{\boldsymbol{k}_0} \sum_{\boldsymbol{k}_0=\boldsymbol{k}_1+\boldsymbol{k}_2} &[V_{\boldsymbol{k}_0}(\boldsymbol{k}_1,\boldsymbol{k}_2)\langle A_{\boldsymbol{k}_0}^* A_{\boldsymbol{k}_1} A_{\boldsymbol{k}_2} \mathrm{e}^{i\Delta\omega_0 t}\rangle \\ &- V_{\boldsymbol{k}_0}^*(\boldsymbol{k}_1,\boldsymbol{k}_2)\langle A_{\boldsymbol{k}_0} A_{\boldsymbol{k}_1}^* A_{\boldsymbol{k}_2}^* \mathrm{e}^{-i\Delta\omega_0 t}\rangle].\end{aligned} \tag{4.109}$$

In this expression, we have switched the dummy variables $\boldsymbol{k}'$, $\boldsymbol{k}''$ to $\boldsymbol{k}_1$ and $\boldsymbol{k}_2$. We also took the time average of the right hand side to eliminate the rapidly changing portion of the products of the amplitudes. If we now make use of the expression (4.101) for the average of the products $A_{\boldsymbol{k}_0}^* A_{\boldsymbol{k}_1} A_{\boldsymbol{k}_2}$ and $A_{\boldsymbol{k}_0} A_{\boldsymbol{k}_1}^* A_{\boldsymbol{k}_2}^*$ in the above equation, there will appear six terms inside the bracket. To evaluate the average of each term, we take one conjugate pair out of the three pairs and designate it by the quantity α:

$$\begin{aligned}\alpha = V_{\boldsymbol{k}_0}(\boldsymbol{k}_1,\boldsymbol{k}_2)&\langle A_{\boldsymbol{k}_0}^* A_{\boldsymbol{k}_1}^{(0)} A_{\boldsymbol{k}_2}^{(0)} \mathrm{e}^{i\Delta\omega_0 t}\rangle \\ - V_{\boldsymbol{k}_0}^*(\boldsymbol{k}_1,\boldsymbol{k}_2)&\langle A_{\boldsymbol{k}_0} A_{\boldsymbol{k}_1}^{(0)*} A_{\boldsymbol{k}_2}^{(0)*} \mathrm{e}^{-i\Delta\omega_0 t}\rangle.\end{aligned} \tag{4.110a}$$

If we then substitute Eq. (4.106),

$$\begin{aligned}\alpha = -S_{\boldsymbol{k}_0} \sum_{\boldsymbol{k}_0=\boldsymbol{k}'+\boldsymbol{k}''} &\left[\frac{|V_{\boldsymbol{k}_0}(\boldsymbol{k}',\boldsymbol{k}'')|^2}{\Delta\omega + i0}\langle A_{\boldsymbol{k}'}^{(0)*} A_{\boldsymbol{k}''}^{(0)*} A_{\boldsymbol{k}_1}^{(0)} A_{\boldsymbol{k}_2}^{(0)}\rangle\right. \\ &\left.- \frac{|V_{\boldsymbol{k}_0}(\boldsymbol{k}',\boldsymbol{k}'')|^2}{\Delta\omega - i0}\langle A_{\boldsymbol{k}'}^{(0)} A_{\boldsymbol{k}''}^{(0)} A_{\boldsymbol{k}_1}^{(0)*} A_{\boldsymbol{k}_2}^{(0)*}\rangle\right].\end{aligned} \tag{4.110b}$$

The average of the products of the four amplitudes can be reduced to sums of products of two amplitudes because of the assumed statistical independence. That is,

$$\begin{aligned}&\langle A_{k'}^{(0)*}A_{k''}^{(0)*}A_{k_1}^{(0)}A_{k_2}^{(0)}\rangle\\&\quad=\langle A_{k'}^{(0)*}A_{k''}^{(0)*}\rangle\langle A_{k_1}^{(0)}A_{k_2}^{(0)}\rangle+\langle A_{k'}^{(0)*}A_{k_1}^{(0)}\rangle\langle A_{k''}^{(0)*}A_{k_2}^{(0)}\rangle\\&\quad+\langle A_{k'}^{(0)*}A_{k_2}^{(0)}\rangle\langle A_{k''}^{(0)*}A_{k_1}^{(0)}\rangle.\end{aligned}\tag{4.111}$$

Now, the average of the products of the two amplitudes which are statistically independent, such as $\langle A_{k'}^{(0)}A_{k''}^{(0)}\rangle$, has a non-vanishing value only when $\boldsymbol{k}'=-\boldsymbol{k}''$. Hence

$$\langle A_{k'}^{(0)}A_{k''}^{(0)}\rangle=|A_{k'}^{(0)}|^2\,\delta_{k',-k''}.\tag{4.112}$$

The use of this relation is called *random phase approximation.* If we note $A_{-k}^{(0)}=A_{k}^{(0)*}$, Eq. (4.111) becomes

$$\begin{aligned}\langle A_{k'}^{(0)*}A_{k''}^{(0)*}A_{k_1}^{(0)}A_{k_2}^{(0)}\rangle&=|A_{k'}^{(0)}|^2\,|A_{k_1}^{(0)}|^2\,\delta_{k',-k''}\delta_{k_1,-k_2}\\&+|A_{k_1}^{(0)}|^2\,|A_{k_2}^{(0)}|^2(\delta_{k_1,k'}\delta_{k_2,k''}+\delta_{k_1,k''}\delta_{k_2,k'}).\end{aligned}\tag{4.113}$$

If $\boldsymbol{k}'=-\boldsymbol{k}''$, $\boldsymbol{k}(=\boldsymbol{k}'+\boldsymbol{k}'')$ becomes zero, hence the first term on the right hand side of this expression does not contribute. We can derive a similar expression for $A_{k'}^{(0)}A_{k''}^{(0)}A_{k_1}^{(0)*}A_{k_2}^{(0)*}$. Substituting these expressions into α, and using the relations

$$\frac{1}{\Delta\omega\pm i0}=P\frac{1}{\Delta\omega}\mp i\pi\,\delta(\Delta\omega),\tag{4.114}$$

$$V_{k_0}(\boldsymbol{k}_1,\boldsymbol{k}_2)=V_{k_0}(\boldsymbol{k}_2,\boldsymbol{k}_1),\tag{4.115}$$

when we perform the summation, we have

$$\alpha=4\pi i S_{k_0}|V_{k_0}(\boldsymbol{k}_1,\boldsymbol{k}_2)|^2\,|A_{k_1}^{(0)}|^2\,|A_{k_2}^{(0)}|^2\,\delta(\Delta\omega_0).\tag{4.116}$$

Similarly, the rest of the conjugate pairs inside the bracket in Eq. (4.109) can be evaluated to give

$$\begin{aligned}&-4\pi i S_{k_1}\,|V_{k_0}(\boldsymbol{k}_1,\boldsymbol{k}_2)|^2\,|A_{k_0}^{(0)}|^2\,|A_{k_2}^{(0)}|^2\,\delta(\Delta\omega_0)\\&-4\pi i S_{k_2}\,|V_{k_0}(\boldsymbol{k}_1,\boldsymbol{k}_2)|^2\,|A_{k_0}^{(0)}|^2\,|A_{k_1}^{(0)}|^2\,\delta(\Delta\omega_0).\end{aligned}$$

If we substitute these results into Eq. (4.109) and remove the superscript (0) which is no longer needed to consider the time development in the τ scale, we finally obtain

$$\frac{\partial |A_{\boldsymbol{k}_0}|^2}{\partial \tau} = 4\pi S_{\boldsymbol{k}_0} \sum_{\boldsymbol{k}_0=\boldsymbol{k}_1+\boldsymbol{k}_2} |V_{\boldsymbol{k}_0}(\boldsymbol{k}_1, \boldsymbol{k}_2)|^2 [S_{\boldsymbol{k}_0} |A_{\boldsymbol{k}_1}|^2 |A_{\boldsymbol{k}_2}|^2 - S_{\boldsymbol{k}_1} |A_{\boldsymbol{k}_0}|^2 |A_{\boldsymbol{k}_2}|^2 - S_{\boldsymbol{k}_2} |A_{\boldsymbol{k}_0}|^2 |A_{\boldsymbol{k}_1}|^2] \, \delta(\omega_{\boldsymbol{k}_0} - \omega_{\boldsymbol{k}_1} - \omega_{\boldsymbol{k}_2}). \tag{4.117}$$

Because of the δ function dependency of the right hand side of the equation, it is convenient to assume a continuous spectrum and use the spectral density function of $|A_{\boldsymbol{k}}|^2$ defined by

$$|A_{\boldsymbol{k}_j}|^2 = J_j(\boldsymbol{k}_j) \, d\boldsymbol{k}_j. \tag{4.118}$$

$J(\boldsymbol{k})$ in this expression is related to the energy spectral density $I(\boldsymbol{k})$ through

$$I(\boldsymbol{k})/\omega = J(\boldsymbol{k}), \tag{4.119}$$

hence $J(\boldsymbol{k})$ may be called the action spectral density. $J(\boldsymbol{k})$ is also related to the spectral density of the number density of wave quantum $N(\boldsymbol{k}) = N_{\boldsymbol{k}}/d\boldsymbol{k}$, by $J(\boldsymbol{k}) = \hbar N(\boldsymbol{k})$.

If we use J, and transform the summation to the integral by noting $\delta_{\boldsymbol{k}_0, \boldsymbol{k}_1+\boldsymbol{k}_2} \to \delta(\boldsymbol{k}_0 - \boldsymbol{k}_1 - \boldsymbol{k}_2)$, we have

$$\frac{\partial J_0(\boldsymbol{k}_0)}{\partial \tau} = 4\pi S_{\boldsymbol{k}_0} \iint d\boldsymbol{k}_1 \, d\boldsymbol{k}_2 \, |V_{\boldsymbol{k}_0}(\boldsymbol{k}_1, \boldsymbol{k}_2)|^2 [S_{\boldsymbol{k}_0} J_1(\boldsymbol{k}_1) J_2(\boldsymbol{k}_2) - S_{\boldsymbol{k}_1} J_0(\boldsymbol{k}_0) J_2(\boldsymbol{k}_2) - S_{\boldsymbol{k}_2} J_0(\boldsymbol{k}_0) J_1(\boldsymbol{k}_1)] \cdot \delta(\boldsymbol{k}_0 - \boldsymbol{k}_1 - \boldsymbol{k}_2) \, \delta(\omega_{\boldsymbol{k}_0} - \omega_{\boldsymbol{k}_1} - \omega_{\boldsymbol{k}_2}). \tag{4.120}$$

Similarly we can obtain equations for $\partial J_1(\boldsymbol{k}_1)/\partial\tau$ and $\partial J_2(\boldsymbol{k}_2)/\partial\tau$ by permuting $0 \to 1$, $1 \to -2$, $2 \to 0$ and $0 \to 2$, $1 \to 0$, $2 \to -1$ respectively and also noting that $J(-\boldsymbol{k}) = J(\boldsymbol{k})$, $S_{-\boldsymbol{k}} = -S_{\boldsymbol{k}}$, and $V_{\boldsymbol{k}_1}(-\boldsymbol{k}_2, \boldsymbol{k}_0) = V_{\boldsymbol{k}_2}(\boldsymbol{k}_0, -\boldsymbol{k}_1) = V_{\boldsymbol{k}_0}(\boldsymbol{k}_1, \boldsymbol{k}_2)$:

$$\frac{\partial J_1(\boldsymbol{k}_1)}{\partial \tau} = -4\pi S_{\boldsymbol{k}_1} \iint d\boldsymbol{k}_2 \, d\boldsymbol{k}_0 \, |V_{\boldsymbol{k}_0}(\boldsymbol{k}_1, \boldsymbol{k}_2)|^2 [S_{\boldsymbol{k}_0} J_1(\boldsymbol{k}_1) J_2(\boldsymbol{k}_2) - S_{\boldsymbol{k}_1} J_0(\boldsymbol{k}_0) J_2(\boldsymbol{k}_2) - S_{\boldsymbol{k}_2} J_0(\boldsymbol{k}_0) J_1(\boldsymbol{k}_1)] \cdot \delta(\boldsymbol{k}_0 - \boldsymbol{k}_1 - \boldsymbol{k}_2) \, \delta(\omega_{\boldsymbol{k}_0} - \omega_{\boldsymbol{k}_1} - \omega_{\boldsymbol{k}_2}), \tag{4.121}$$

$$\frac{\partial J_2(\boldsymbol{k}_2)}{\partial \tau} = -4\pi S_{\boldsymbol{k}_2} \iint d\boldsymbol{k}_1 \, d\boldsymbol{k}_0 \, |V_{\boldsymbol{k}_0}(\boldsymbol{k}_1, \boldsymbol{k}_2)|^2 [S_{\boldsymbol{k}_0} J_1(\boldsymbol{k}_1) J_2(\boldsymbol{k}_2) - S_{\boldsymbol{k}_1} J_0(\boldsymbol{k}_0) J_2(\boldsymbol{k}_2) - S_{\boldsymbol{k}_2} J_0(\boldsymbol{k}_0) J_1(\boldsymbol{k}_1)] \cdot \delta(\boldsymbol{k}_0 - \boldsymbol{k}_1 - \boldsymbol{k}_2) \, \delta(\omega_{\boldsymbol{k}_0} - \omega_{\boldsymbol{k}_1} - \omega_{\boldsymbol{k}_2}). \tag{4.122}$$

Eqs. (4.120) to (4.122) correspond to the Eqs. (4.88) for coherent three-wave interactions. If we compare these two sets of coupled equations, we can see interesting contrasts between the coherent three-wave interactions and the present case of random phase.

First, in the present case, the coupling time scale is proportional to the energy of the wave, while, in the former case, it was proportional to the amplitude. Consequently the nonlinear development in the random phase system is much slower than in the coherent phase system, as may be expected.

Another important difference is that the process described by the coupled Eqs. (4.120) to (4.122) is *irreversible* while as we have seen the coherent three-wave interaction is *reversible*. To prove this aspect, let us calculate the time rate of change of the nonequilibrium entropy of the system described by Eqs. (4.120) to (4.122). The entropy $S(\tau)$ here may be defined by

$$S(\tau)=\sum_{j=0,1,2}\int \ln J_j(\boldsymbol{k}_j)\,d\boldsymbol{k}_j. \tag{4.123}$$

If we substitute Eqs. (4.120), (4.121) and (4.122) into this expression, we have

$$\begin{aligned}\frac{\partial S}{\partial \tau}&=\sum_j \int d\boldsymbol{k}_j \frac{1}{J_j}\frac{\partial J_j}{\partial \tau}\\ &=4\pi \iiint d\boldsymbol{k}_0\, d\boldsymbol{k}_1\, d\boldsymbol{k}_2\, |V_{\boldsymbol{k}_0}(\boldsymbol{k}_1,\boldsymbol{k}_2)|^2 J_0(\boldsymbol{k}_0) J_1(\boldsymbol{k}_1) J_2(\boldsymbol{k}_2)\\ &\quad\cdot\left[\frac{S_{\boldsymbol{k}_0}}{J_0(\boldsymbol{k}_0)}-\frac{S_{\boldsymbol{k}_1}}{J_1(\boldsymbol{k}_1)}-\frac{S_{\boldsymbol{k}_2}}{J_2(\boldsymbol{k}_2)}\right]^2\\ &\quad\cdot\delta(\boldsymbol{k}_0-\boldsymbol{k}_1-\boldsymbol{k}_2)\,\delta(\omega_{\boldsymbol{k}_0}-\omega_{\boldsymbol{k}_1}-\omega_{\boldsymbol{k}_2})\geqq 0.\end{aligned} \tag{4.124}$$

Thus the entropy is a monotonically increasing function of time and hence the process is irreversible. The quasistationary condition is reached when

$$\frac{S_{\boldsymbol{k}_0}}{J_0(\boldsymbol{k}_0)}=\frac{S_{\boldsymbol{k}_1}}{J_1(\boldsymbol{k}_1)}+\frac{S_{\boldsymbol{k}_2}}{J_2(\boldsymbol{k}_2)} \tag{4.125}$$

for each set of waves satisfying $\boldsymbol{k}_0=\boldsymbol{k}_1+\boldsymbol{k}_2$ and $\omega_{\boldsymbol{k}_0}=\omega_{\boldsymbol{k}_1}+\omega_{\boldsymbol{k}_2}$.

To study some concrete example of this irreversible decay process, we consider a one dimensional problem in which a wave in a wave packet at $\omega\sim\omega_{\boldsymbol{k}_0}$ decays into two other waves in wave packets at $\omega\sim\omega_{\boldsymbol{k}_1}$ and $\omega_{\boldsymbol{k}_2}$. In such a case, the integrations in Eqs. (4.120) to (4.122) can readily be performed (Davidson and Kaufman, 1969) to give

$$\begin{aligned}
\frac{1}{\alpha_0}\frac{\partial J_0}{\partial \tau} &= (J_1 J_2 - J_0 J_2 - J_0 J_1) \\
\frac{1}{\alpha_1}\frac{\partial J_1}{\partial \tau} &= -(J_1 J_2 - J_0 J_2 - J_0 J_1) \\
\frac{1}{\alpha_2}\frac{\partial J_2}{\partial \tau} &= -(J_1 J_2 - J_0 J_2 - J_0 J_1).
\end{aligned} \tag{4.126}$$

In these expressions all J_j's are functions of k_0; k_1 and k_2 have been eliminated by the conditions

$$k_0 = k_1 + k_2 \quad \text{and} \quad \omega_{k_0} = \omega_{k_1} + \omega_{k_2}$$

$\alpha_0, \alpha_1, \alpha_2$ are given by

$$\begin{aligned}
\alpha_0 &= 4\pi\, |V_{k_0}(k_1, k_2)|^2 \left| \frac{d\omega_{k_1}}{dk_1} - \frac{d\omega_{k_2}}{dk_2} \right|^{-1} \\
\alpha_1 &= 4\pi\, |V_{k_0}(k_1, k_2)|^2 \left| \frac{d\omega_{k_2}}{dk_2} - \frac{d\omega_{k_0}}{dk_0} \right|^{-1} \\
\alpha_2 &= 4\pi\, |V_{k_0}(k_1, k_2)|^2 \left| \frac{d\omega_{k_0}}{dk_0} - \frac{d\omega_{k_1}}{dk_1} \right|^{-1}.
\end{aligned} \tag{4.127}$$

We have also assumed that all three waves are positive energy waves so that $S_{k_0}, S_{k_1}, S_{k_2} > 0$.

Let us consider a situation very early in time when only the ω_{k_0} mode is populated, thus $J_1(-\infty) = J_2(-\infty) \simeq 0$. By adding each two pairs of Eqs. (4.126), we obtain the following constants of motion:

$$\begin{aligned}
\frac{J_0}{\alpha_0} + \frac{J_1}{\alpha_1} &= C_1 \\
\frac{J_0}{\alpha_0} + \frac{J_2}{\alpha_2} &= C_2.
\end{aligned} \tag{4.128}$$

From the assumed initial conditions, we have

$$C_1 = C_2 = J_0(-\infty)/\alpha_0. \tag{4.129}$$

Substituting Eqs. (4.128) and (4.129) into the first equation of Eqs. (4.126), we have

$$\frac{\partial J_0}{\partial \tau}=\frac{\alpha_1\alpha_2}{\alpha_0}\frac{J_0(-\infty)}{J_0(\infty)}[J_0(\tau)-J_0(-\infty)][J_0(\tau)-J_0(\infty)] \quad (4.130)$$

where

$$J_0(\infty)=J_0(-\infty)\left[1+\frac{\alpha_0(\alpha_1+\alpha_2)}{\alpha_1\alpha_2}\right]^{-1}\leq J_0(-\infty).$$

Because all of J_j's must be positive, from Eq. (4.130), $J_0(\tau)<J_0(-\infty)$. Also, it can be shown easily that $J_0(\infty)$ corresponds to the condition of maximum entropy (Eq. (4.125)), hence $\partial J_0/\partial\tau\leq 0$ and J_0 is bound between the two values of J_0; $J_0(-\infty)$ and $J_0(\infty)$. Eq. (4.130) can be easily integrated to give

$$J_0(\tau)=J_0(-\infty)-\frac{J_0(-\infty)-J_0(\infty)}{2}\left(1+\tanh\frac{\alpha\tau}{2}\right) \quad (4.131)$$

where

$$\alpha=[J(-\infty)-J(\infty)]\frac{\alpha_1\alpha_2 J_0(-\infty)}{\alpha_0 J_0(\infty)}=(\alpha_1+\alpha_2)J_0(-\infty). \quad (4.132)$$

Corresponding solutions for J_1 and J_2 are obtained from Eqs. (4.128) and (4.129):

$$J_1(\tau)=\frac{\alpha_1}{\alpha_2}[J_0(-\infty)-J_0(\tau)]=\frac{\alpha_1}{\alpha_0}\frac{J_0(-\infty)-J_0(\infty)}{2}\left(1+\tanh\frac{\alpha\tau}{2}\right)$$

$$J_2(\tau)=\frac{\alpha_2}{\alpha_0}\frac{J_0(-\infty)-J_0(\infty)}{2}\left(1+\tanh\frac{\alpha\tau}{2}\right).$$

The behavior of $J_j(\tau)$ is shown in Fig. 42. In the time coordinate, all the solutions have shock structures. The time constant $(\alpha/2)^{-1}$ is given by Eq. (4.132) and can be seen to be proportional to $J_0(-\infty)^{-1}$. Specifically, for a given strength of the coupling coefficient $(\alpha_1+\alpha_2)$, the decay time constant is inversely proportional to the population density at $t\to-\infty$. It can also be seen from these solutions, that $J_j(\infty)$ does in fact satisfy the state of maximum entropy as shown in Eq. (4.125).

If we take the ω_{k_0} mode to be linearly unstable, the saturation process due to the three wave decay can be studied by adding a term $\gamma_0 J_0$ on the right hand side of the first equation in (4.126). To achieve the saturation, the two modes at $\omega\sim\omega_{k_1}$ and ω_{k_2} must be linearly stable with finite dampings. In particular, if we assume the mode at $\omega\sim\omega_{k_2}$ to be heavily damped so that its amplitude can never grow to an appreciable level,

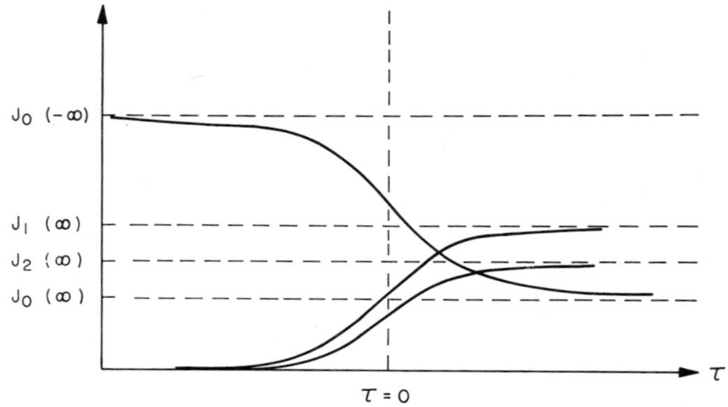

Fig. 42. Temporal behavior of spectral density of actions $J_j(k)$ in nonlinear three wave interactions with random phase

then the coupled Eqs. (4.126) reduce to coupled equations for two waves given by

$$\begin{aligned} \frac{1}{\alpha_0}\left(\frac{\partial J_0}{\partial \tau}-\gamma_0 J_0\right) &= -J_0 J_1 \\ \frac{1}{\alpha_1}\left(\frac{\partial J_1}{\partial \tau}+\gamma_1 J_1\right) &= J_0 J_1 \end{aligned} \tag{4.133}$$

where γ_0 and γ_1 are the growth and the damping rates of J_0 and J_1 respectively.

At an early stage of the linear instability, J_0 will grow exponentially. When J_0 becomes larger than γ_1/α_1, the linear damping rate of J_1, J_1 will shoot up suddenly at a rate faster than exponential and will almost immediately stabilize the linear instability of J_0 through the coupling term $J_0 J_1$ in the first equation of (4.133). This type of relaxation process will continue until interactions with other waves diffuse the spectrum. The coupled Eq. (4.133) is similar to the case of the saturation due to nonlinear Landau damping. A numerical evaluation of this type of equation will be shown in Subsection 4.3d.

Saturation of $\boldsymbol{E} \times \boldsymbol{B}$ instability in the ionospheric plasma due to spectral diffusion in angular directions of $\boldsymbol{k}$ vectors through nonlinear mode-mode coupling has been discussed recently by Rogister (1973).

So far we have assumed that all three waves are positive energy waves. If, however, the $\omega_{\boldsymbol{k}_0}$ mode is a negative energy mode, an explosive instability occurs (as was the case with coherent three-wave interactions). As we have seen in Subsection 2.5b, waves in the presence of a density gradient *and* an electron drift in the partially ionized plasma have a property similar to that of a negative energy wave. Sato (1971) has shown in fact, that the simultaneous presence of a density gradient and an electron $\boldsymbol{E}\times\boldsymbol{B}$ drift in the ionosphere leads to explosive mode coupling. Consequently, the use of nonlinear mode coupling to obtain a stabilization mechanism requires careful treatment.

4.3d Nonlinear Two Wave Interactions Due to Scattering by Particles – Nonlinear Landau Damping

As we have seen briefly at the end of the previous subsection, the nonlinear three wave interaction does exist even if the third wave is a heavily damped mode (quasi-mode). We consider here a case in which, instead of a heavily damped third mode, a nonlinear interaction with particles is involved. When a wave with wave number $\boldsymbol{k}$ and frequency $\omega_{\boldsymbol{k}}$ is scattered by a particle with energy $\mathscr{E}$ and momentum $\boldsymbol{p}$ and is transformed into another wave with wave number $\boldsymbol{k}'$ and frequency $\omega_{\boldsymbol{k}'}$, the momentum and energy conservation laws give

$$\hbar\boldsymbol{k}'+\boldsymbol{p}'=\hbar\boldsymbol{k}+\boldsymbol{p} \tag{4.134}$$

$$\hbar\omega_{\boldsymbol{k}'}+\mathscr{E}(\boldsymbol{p}')=\hbar\omega_{\boldsymbol{k}}+\mathscr{E}(\boldsymbol{p}), \tag{4.135}$$

where $\boldsymbol{p}'$ represents the particle momentum after the scattering. Because the wave energy $\hbar\omega$ and momentum $\hbar\boldsymbol{k}$ are much smaller than the corresponding values for the particle, we can expand $\mathscr{E}(\boldsymbol{p}')$ around $\mathscr{E}(\boldsymbol{p})$ to give

$$\begin{aligned}\mathscr{E}(\boldsymbol{p}')&=\mathscr{E}[\boldsymbol{p}+\hbar(\boldsymbol{k}-\boldsymbol{k}')]\\&\simeq\mathscr{E}(\boldsymbol{p})+\hbar(\boldsymbol{k}-\boldsymbol{k}')\cdot\frac{\partial\mathscr{E}}{\partial\boldsymbol{p}}\\&=\mathscr{E}(\boldsymbol{p})+\hbar(\boldsymbol{k}-\boldsymbol{k}')\cdot\boldsymbol{v}\\&=\mathscr{E}(\boldsymbol{p})+\hbar(\omega_{\boldsymbol{k}}-\omega_{\boldsymbol{k}'}).\end{aligned} \tag{4.136}$$

Hence the resonant condition in this process is given by

$$\omega_{\boldsymbol{k}}-\omega_{\boldsymbol{k}'}=(\boldsymbol{k}-\boldsymbol{k}')\cdot\boldsymbol{v}. \tag{4.137}$$

Through this process the wave at $\omega_{\boldsymbol{k}}$ transfers its energy into another wave $\omega_{\boldsymbol{k}'}$ and a particle that satisfies the resonant condition. Hence this process works to diffuse energy from the linearly unstable modes to stable modes, and thereby the instability may be saturated.

We can obtain a wave kinetic equation to describe the dynamic change in wave amplitude by using the perturbation expansion of the Vlasov equation as in the case of nonlinear wave-wave interactions. However, before we do the detailed calculation, let us examine the basic properties of this process. If we use the particle picture of a wave, a wave-particle interaction may be considered as a collision between wave quanta and particles. For example, linear Landau damping can be looked at from such a point of view. The particle gains momentum $\hbar\boldsymbol{k}$ from the interaction, hence the total *loss* of the number density of waves $N_{\boldsymbol{k}}$ per unit time in such a collisional process must be given by

$$\begin{aligned} -\int f_0(\boldsymbol{p}+\hbar\boldsymbol{k})\, N_{\boldsymbol{k}}\, w(\boldsymbol{k},\boldsymbol{p})\, d\boldsymbol{p} + \int f_0(\boldsymbol{p})\, N_{\boldsymbol{k}}\, w(\boldsymbol{k},\boldsymbol{p})\, d\boldsymbol{p} \\ = -N_{\boldsymbol{k}} \int w(\boldsymbol{k},\boldsymbol{p})\, \hbar\boldsymbol{k}\cdot\frac{\partial f_0}{\partial \boldsymbol{p}}\, d\boldsymbol{p} \end{aligned} \tag{4.138}$$

where $f_0(\boldsymbol{p})$ is the particle distribution function and $w(\boldsymbol{k},\boldsymbol{p})$ designates the transition probability of this process, which is effectively the cross section of the collision. Here as before, we assumed $\boldsymbol{p} \gg \hbar\boldsymbol{k}$. If we equate this loss rate with $-\partial N_{\boldsymbol{k}}/\partial t$, we have

$$\frac{\partial N_{\boldsymbol{k}}}{\partial t} = N_{\boldsymbol{k}} \int w(\boldsymbol{k},\boldsymbol{p})\, \hbar\boldsymbol{k}\cdot\frac{\partial f_0}{\partial \boldsymbol{p}}\, d\boldsymbol{p} \equiv 2\gamma_{\boldsymbol{k}} N_{\boldsymbol{k}}; \tag{4.139}$$

$(-\gamma_{\boldsymbol{k}})$ should correspond with the linear Landau damping rate given by (4.15),

$$2\gamma_{\boldsymbol{k}} = \frac{\pi\omega_{pe}^3}{k^2} \int \delta(\boldsymbol{k}\cdot\boldsymbol{v}-\omega_{\boldsymbol{k}})\, \boldsymbol{k}\cdot\frac{\partial f_0}{\partial \boldsymbol{v}}\, d\boldsymbol{v}. \tag{4.140}$$

Comparing these two expressions, we have

$$w(\boldsymbol{k},\boldsymbol{p}) = \frac{\pi\omega_{pe}^3}{\hbar k^2}\, \delta(\boldsymbol{k}\cdot\boldsymbol{v}-\omega_{\boldsymbol{k}}), \tag{4.141}$$

which is nothing but the rate of Cherenkov emission of plasmons by particles with density n_0 and velocity $\boldsymbol{v}$. (Note that the Cherenkov emission rate of a plasma wave by a single particle is given by $\pi e^2 \omega_{pe} (\hbar k^2 m_e \varepsilon_0)^{-1} \delta(\boldsymbol{k}\cdot\boldsymbol{v}-\omega_{\boldsymbol{k}})$.)

We now consider a nonlinear interaction of two waves and a particle. In this case, the particle increases its momentum by an amount $\hbar(\boldsymbol{k}-\boldsymbol{k}')$ and the transfer process must be proportional to $N_{\boldsymbol{k}} N_{\boldsymbol{k}'} f_0$. Hence we have, instead of (4.139),

$$\frac{\partial N_{\boldsymbol{k}}}{\partial t} = N_{\boldsymbol{k}} \sum_{\boldsymbol{k}'} \int d\boldsymbol{p}\, w_{\boldsymbol{k}}(\boldsymbol{k}',\boldsymbol{p})\, N_{\boldsymbol{k}'}\, \hbar(\boldsymbol{k}-\boldsymbol{k}')\cdot\frac{\partial f_0}{\partial \boldsymbol{p}} \tag{4.142a}$$

where $w_{\boldsymbol{k}}(\boldsymbol{k}', \boldsymbol{p})$ is the probability of the scattering. If we rewrite the derivative of the particle distribution function f_0 with respect to momentum as that with respect to energy $\mathscr{E}$, because $\partial\mathscr{E}/\partial\boldsymbol{p}=\boldsymbol{v}$, Eq. (4.142a) becomes

$$\begin{aligned}\frac{\partial N_{\boldsymbol{k}}}{\partial t} &= N_{\boldsymbol{k}} \sum_{\boldsymbol{k}'} \int d\boldsymbol{p}\, w_{\boldsymbol{k}}(\boldsymbol{k}', \boldsymbol{p})\, N_{\boldsymbol{k}'}\, \hbar(\boldsymbol{k}-\boldsymbol{k}')\cdot\boldsymbol{v}\frac{\partial f_0}{\partial\mathscr{E}} \\ &= N_{\boldsymbol{k}} \sum_{\boldsymbol{k}'} \int d\boldsymbol{p}\, w_{\boldsymbol{k}}(\boldsymbol{k}', \boldsymbol{p})\, N_{\boldsymbol{k}'}\, \hbar(\omega_{\boldsymbol{k}}-\omega_{\boldsymbol{k}'})\frac{\partial f_0}{\partial\mathscr{E}}.\end{aligned} \tag{4.142b}$$

If the plasma is in thermodynamical equilibrium, $\partial f_0/\partial\mathscr{E}<0$, hence $\partial N_{\boldsymbol{k}}/\partial t<0$ (>0) if $\omega_{\boldsymbol{k}}>\omega_{\boldsymbol{k}'}$ $(\omega_{\boldsymbol{k}}<\omega_{\boldsymbol{k}'})$. Hence nonlinear Landau damping transforms the number of waves at the higher frequency to the lower frequency. Because the probability of scattering is symmetric, i.e., $w_{\boldsymbol{k}}(\boldsymbol{k}', \boldsymbol{p})=w_{\boldsymbol{k}'}(\boldsymbol{k}, \boldsymbol{p})$, we can also see that the total number of waves, $\sum_{\boldsymbol{k}} N_{\boldsymbol{k}}$, is conserved by this process. However, the total energy of the waves is not conserved and the difference is transferred into the particle energy.

As was the case with linear Landau damping, to calculate $w_{\boldsymbol{k}}(\boldsymbol{k}', \boldsymbol{p})$, we have to go back to the kinetic equation and calculate for each case the different waves involved. Let us take again the example of electrostatic modes. As we saw in Subsection 4.3b, Eq. (4.75), the nonlinear dispersion relation for an electrostatic mode may be written as

$$\begin{aligned}&\varepsilon_{\boldsymbol{k}}^{(1)}\varphi_{\boldsymbol{k}} + \sum_{\boldsymbol{k}=\boldsymbol{k}'+\boldsymbol{k}''} \varepsilon_{\boldsymbol{k}}^{(2)}(\boldsymbol{k}', \boldsymbol{k}'')\,\varphi_{\boldsymbol{k}'}\varphi_{\boldsymbol{k}''} \\ &\quad + \sum_{\boldsymbol{k}=\boldsymbol{k}'+\boldsymbol{k}''+\boldsymbol{k}'''} \varepsilon_{\boldsymbol{k}}^{(3)}(\boldsymbol{k}', \boldsymbol{k}'', \boldsymbol{k}''')\,\varphi_{\boldsymbol{k}'}\varphi_{\boldsymbol{k}''}\varphi_{\boldsymbol{k}'''} + \cdots = 0\end{aligned} \tag{4.75}$$

in which $\varepsilon^{(1)}$, $\varepsilon^{(2)}$, $\varepsilon^{(3)}$ are given by Eqs. (4.76), (4.77) and (4.78) respectively for the case without an external magnetic field.

Let us construct the wave kinetic equation associated with the nonlinear Landau damping from the above coupled equation. We consider a coupling between $\boldsymbol{k}$ mode and $\boldsymbol{k}'$ mode through a beat mode $\boldsymbol{k}-\boldsymbol{k}'$ which is *not* a linearly resonant mode of the system, i.e., $\varepsilon_{\boldsymbol{k}-\boldsymbol{k}'}\neq 0$ for a real frequency. We assume, however, that both the $\boldsymbol{k}$ and $\boldsymbol{k}'$ modes are linearly resonant modes.

First we consider only to the order of $\varepsilon^{(2)}$ and write down the equations for $\varphi_{\boldsymbol{k}}$,

$$\varepsilon_{\boldsymbol{k}}^{(1)}\varphi_{\boldsymbol{k}} + \sum_{\boldsymbol{k}'} \varepsilon_{\boldsymbol{k}}^{(2)}(\boldsymbol{k}', \boldsymbol{k}-\boldsymbol{k}')\,\varphi_{\boldsymbol{k}'}\varphi_{\boldsymbol{k}-\boldsymbol{k}'} = 0. \tag{4.143}$$

Now, for $\phi_{\boldsymbol{k}-\boldsymbol{k}'}$ mode we have

$$\varepsilon_{\boldsymbol{k}-\boldsymbol{k}'}^{(1)}\varphi_{\boldsymbol{k}-\boldsymbol{k}'} + \sum_{\boldsymbol{k}''} \varepsilon_{\boldsymbol{k}-\boldsymbol{k}'}^{(2)}(\boldsymbol{k}'', \boldsymbol{k}-\boldsymbol{k}'-\boldsymbol{k}'')\,\varphi_{\boldsymbol{k}''}\varphi_{\boldsymbol{k}-\boldsymbol{k}'-\boldsymbol{k}''} = 0,$$

where, if we put $\boldsymbol{k}''=\boldsymbol{k}$,

$$\varepsilon^{(1)}_{\boldsymbol{k}-\boldsymbol{k}'}\,\varphi_{\boldsymbol{k}-\boldsymbol{k}'}+\varepsilon^{(2)}_{\boldsymbol{k}-\boldsymbol{k}'}(\boldsymbol{k},\,-\boldsymbol{k}')\,\varphi_{\boldsymbol{k}}\,\varphi_{-\boldsymbol{k}'}=0. \tag{4.144}$$

If we substitute (4.144) into (4.143), we have

$$\varepsilon^{(1)}_{\boldsymbol{k}}\,\varphi_{\boldsymbol{k}}=\sum_{\boldsymbol{k}'}\frac{\varepsilon^{(2)}_{\boldsymbol{k}}(\boldsymbol{k}',\boldsymbol{k}-\boldsymbol{k}')\,\varepsilon^{(2)}_{\boldsymbol{k}-\boldsymbol{k}'}(\boldsymbol{k},\,-\boldsymbol{k}')}{\varepsilon^{(1)}_{\boldsymbol{k}-\boldsymbol{k}'}}\,|\varphi_{\boldsymbol{k}'}|^2\,\varphi_{\boldsymbol{k}}.$$

If we permute $\boldsymbol{k}'$ with $\boldsymbol{k}-\boldsymbol{k}'$ as well as $\boldsymbol{k}$ and $-\boldsymbol{k}'$, we have four of these terms. Similarly, from $\varepsilon^{(3)}$, there are three of the $\varepsilon^{(3)}_{\boldsymbol{k}}(\boldsymbol{k}',\boldsymbol{k},\,-\boldsymbol{k}')$ terms. Thus combining all of these, we have

$$\begin{aligned}&\varepsilon^{(1)}_{\boldsymbol{k}}\,\varphi_{\boldsymbol{k}}\\&=\sum_{\boldsymbol{k}'}\left[\frac{4\varepsilon^{(2)}_{\boldsymbol{k}}(\boldsymbol{k}',\boldsymbol{k}-\boldsymbol{k}')\,\varepsilon^{(2)}_{\boldsymbol{k}-\boldsymbol{k}'}(\boldsymbol{k},\,-\boldsymbol{k}')}{\varepsilon^{(1)}_{\boldsymbol{k}-\boldsymbol{k}'}}-3\varepsilon^{(3)}_{\boldsymbol{k}}(\boldsymbol{k}',\boldsymbol{k},\,-\boldsymbol{k}')\right]|\varphi_{\boldsymbol{k}'}|^2\,\varphi_{\boldsymbol{k}}.\end{aligned} \tag{4.145}$$

Note that because the beat mode is non-resonant ($\varepsilon^{(1)}_{\boldsymbol{k}-\boldsymbol{k}'}\neq 0$), we must retain $\varepsilon^{(3)}_{\boldsymbol{k}}$ contribution here. As we saw in Eq. (4.79), $\varepsilon^{(1)}_{\boldsymbol{k}}\varphi_{\boldsymbol{k}}$ may be written as

$$\varepsilon^{(1)}_{\boldsymbol{k}}\,\varphi_{\boldsymbol{k}}=i\,\frac{\partial\varepsilon}{\partial\omega}\left[-\gamma_{\boldsymbol{k}}+\frac{\partial}{\partial\tau}+\boldsymbol{v}_g\cdot\frac{\partial}{\partial\xi}\right]\varphi_{\boldsymbol{k}}. \tag{4.79}$$

Hence, if we ignore spatial variation as well as the linear damping (growth) rate $\gamma_{\boldsymbol{k}}$, we have from (4.145) and (4.79)

$$\begin{aligned}\left.\frac{\partial\varepsilon}{\partial\omega}\right|_{\boldsymbol{k}}\frac{\partial\varphi_{\boldsymbol{k}}}{\partial\tau}=\mathrm{Im}\sum_{\boldsymbol{k}'}&\left[\frac{4\varepsilon^{(2)}_{\boldsymbol{k}}(\boldsymbol{k}',\boldsymbol{k}-\boldsymbol{k}')\,\varepsilon^{(2)}_{\boldsymbol{k}-\boldsymbol{k}'}(\boldsymbol{k},\,-\boldsymbol{k}')}{\varepsilon^{(1)}_{\boldsymbol{k}-\boldsymbol{k}'}}\right.\\&\left.-3\,\varepsilon^{(3)}_{\boldsymbol{k}}(\boldsymbol{k}',\boldsymbol{k},\,-\boldsymbol{k}')\right]|\varphi_{\boldsymbol{k}'}|^2\,\varphi_{\boldsymbol{k}}.\end{aligned} \tag{4.146}$$

Furthermore if we multiply both sides by $\varphi^*_{\boldsymbol{k}}$ and construct $N_{\boldsymbol{k}}$ from $|\varphi_{\boldsymbol{k}}|^2\left|\frac{k^2}{2\hbar\omega_{\boldsymbol{k}}}\,\frac{\partial(\omega\varepsilon_0\varepsilon)}{\partial\omega}\right|^2$, we can see that Eq. (4.146) has a similar structure to Eq. (4.142b), although it is not obvious from the above expression whether the dependency on f_0 is of the type shown in Eq. (4.142b). To reduce this expression to the form of Eq. (4.142b), and hence to obtain the explicit form of $w_{\boldsymbol{k}}(\boldsymbol{k}',\boldsymbol{p})$, is rather tedious (see for example, Nishikawa, 1970).

Let us now investigate the process by which the nonlinear Landau damping quenches the linear instability. If we use the symmetry property of the coupling coefficient $w_{\boldsymbol{k}}(\boldsymbol{k}', \boldsymbol{p})$, the coupled equation relating the actions I_0 and I_1 of the two waves may be written as

$$\frac{dI_0}{d\tau} = \gamma_0 I_0 - \alpha I_0 I_1 + R, \tag{4.147}$$

$$\frac{dI_1}{d\tau} = -\gamma_1 I_1 + \alpha I_0 I_1 + S \tag{4.148}$$

where α is the coupling coefficient given by

$$\alpha = -\int d\boldsymbol{p}\, w_{\boldsymbol{k}_0}(\boldsymbol{k}_1, \boldsymbol{p})\hbar(\omega_{\boldsymbol{k}_0} - \omega_{\boldsymbol{k}_1})\frac{\partial f_0}{\partial \mathscr{E}}. \tag{4.149}$$

Here $\omega_{\boldsymbol{k}_1}$ is assumed to be smaller than $\omega_{\boldsymbol{k}_0}$, γ_0 and γ_1 are the linear growth and damping rates of I_0 and I_1 respectively, and R and S are small but finite source terms representing spontaneous emission rates.

In the absence of the sources, the above set of equations can be integrated easily to give

$$\alpha I_0 - \gamma_1 \ln I_0 = \gamma_0 \ln I_1 - \alpha I_1 + c. \tag{4.150}$$

In the $I_0 - I_1$ plane, this expression becomes a closed loop, indicating a bound, and periodic motion. The qualitative behaviors of I_0 and I_1 as functions of time are plotted in Fig. 43a.

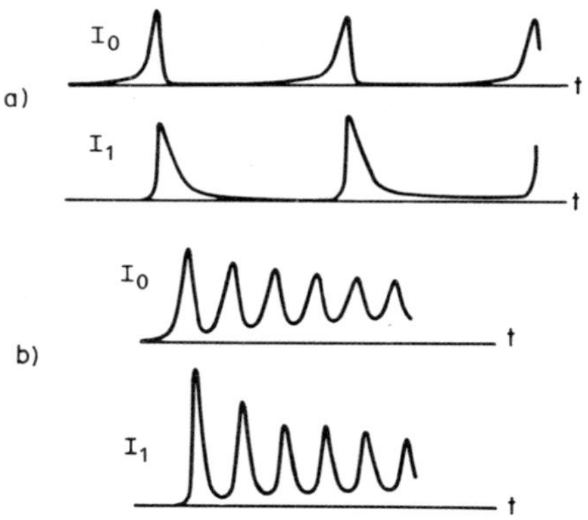

Fig. 43a and b. Temporal behavior of intensities of two waves, one linearly unstable (I_0), and one stable (I_1), coupled through a nonlinear interaction with particles (nonlinear Landau damping); (a) in the absence of source terms and (b) in the presence of the source terms. (After White *et al.*, 1972)

White *et al.* (1972), however, have pointed out the importance of the emission terms R and S in the dynamics of I_0 and I_1. They have shown that even if these quantities are small, their presence destroys the recurrence nature of I_0 and I_1 and the temporal oscillation damps out as shown in Fig. 34b.

Let us now consider effects of the convection term on nonlinear Landau damping. If we ignore the linear growth and damping rates and also the source terms in Eqs. (4.147) and (4.148), but introduce the convective effect on the left hand side, we have

$$\begin{aligned} \frac{\partial I_0}{\partial \tau} + v_0 \frac{\partial I_0}{\partial \xi} &= -\alpha I_0 I_1 \\ \frac{\partial I_1}{\partial \tau} + v_1 \frac{\partial I_1}{\partial \xi} &= \alpha I_0 I_1, \end{aligned} \tag{4.151}$$

where v_0 and v_1 are the group velocities of 0 and 1 waves respectively.

We seek a stationary solution with respect to a coordinate moving at a speed λ. For this we introduce the coordinate $X = \xi - \lambda\tau$. Then Eqs. (4.151) become

$$\begin{aligned} (v_0 - \lambda) \frac{dI_0}{dX} &= -\alpha I_0 I_1 \\ (v_1 - \lambda) \frac{dI_1}{dX} &= \alpha I_0 I_1. \end{aligned} \tag{4.152}$$

The constant of motion is immediately obtained by adding these equations,

$$(v_0 - \lambda) I_0 + (v_1 - \lambda) I_1 = c. \tag{4.153}$$

The solution of the coupled equations can then be expressed by

$$I_0 = \frac{c}{2(v_0 - \lambda)} - \frac{1}{2} \frac{|c|}{|v_0 - \lambda|} \operatorname{Sign}(v_1 - \lambda) \tanh \gamma X, \tag{4.154}$$

$$I_1 = \frac{c}{2(v_1 - \lambda)} + \frac{1}{2} \frac{|c|}{|v_1 - \lambda|} \operatorname{Sign}(v_0 - \lambda) \tanh \gamma X \tag{4.155}$$

where

$$\gamma = \frac{1}{2} \frac{\alpha |c|}{|(v_1 - \lambda)(v_0 - \lambda)|}. \tag{4.156}$$

Because I_0 and I_1 must be positive, from the solutions (4.154) and (4.155), we see:

(a) if $c>0$

$$v_0-\lambda>0 \quad \text{and} \quad v_1-\lambda>0,$$

(b) if $c<0$

$$v_0-\lambda<0 \quad \text{and} \quad v_1-\lambda<0.$$

Consequently, if $\lambda>0$, I_0 becomes a shock wave, while I_1 is a rarefaction wave with respect to the ξ coordinate, with the shock thickness given by γ^{-1}. The stationary speed λ in the interaction may be given as a function of the value of actions I_0 and I_1 as $X\to\mp\infty$ respectively. If we write $I_0(-\infty)=\bar{I}_0$ and $I_1(+\infty)=\bar{I}_1$, then from Eqs. (4.154) and (4.155),

$$\bar{I}_0=\frac{c}{v_0-\lambda}, \tag{4.157}$$

$$\bar{I}_1=\frac{c}{v_1-\lambda} \tag{4.158}$$

hence

$$\lambda=\frac{v_0\bar{I}_0-v_1\bar{I}_1}{\bar{I}_0-\bar{I}_1}. \tag{4.159}$$

Note that when the condition $v_0\bar{I}_0=v_1\bar{I}_1$ is chosen, λ becomes zero; this gives a standing shock wave and a standing rarefaction wave.

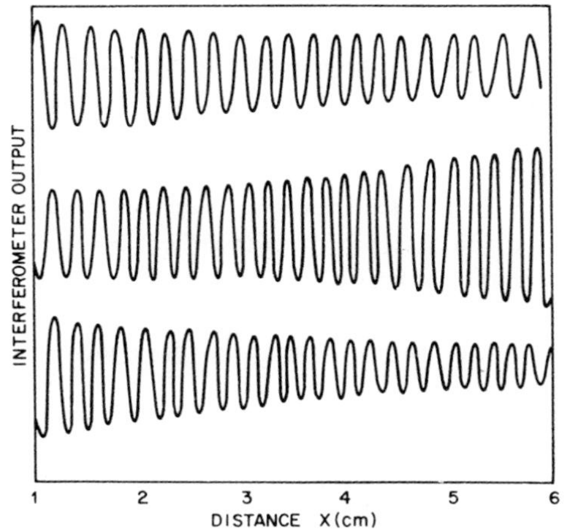

Fig. 44. Interferometer traces showing nonlinear growth and damping of the ion acoustic wave at $\omega_1/2\pi=0.9$ MHz. The top trace is with no ω_2 wave, the middle with the ω_2 wave at $\omega_2/2\pi=1.0$ MHz, the bottom trace with ω_2 wave at $\omega_2/2\pi=0.8$ MHz, in which $\omega_{pi}/2\pi=$ $=1.3$ MHz. (After Ikezi and Kiwamoto, 1971)

Although this final result is not surprising, it is interesting to note that the shock speed can be controlled by combination of the parameters v_0, v_1, $\bar{I}_0$ and $\bar{I}_1$.

Application of nonlinear Landau damping to the observed diffuse plasma resonances at harmonics of electron cyclotron frequencies in the ionosphere was made by Oya (1971). In laboratory plasma, nonlinear Landau damping was observed for example in one-dimensional ion acoustic waves due to scattering by ions (Ikezi and Kiwamoto, 1971). Since the group velocity $\partial\omega/\partial k$ is smaller at $\omega \lesssim \omega_{pi}$ (ion plasma frequency), the beat wave of two waves with frequencies ω_1 and ω_2 ($\omega_1 > \omega_2$) close to ω_{pi} is nonresonant, i.e., $\varepsilon^{(1)}_{k_1-k_2}(\omega_1-\omega_2) \neq 0$; hence the decay instability is prohibited. However, if $(\omega_1-\omega_2)/(k_1-k_2)$ is close to the ion thermal speed, the ω_1 wave is effectively scattered into the ω_2 wave by the ions.

Fig. 44 shows the output of an interferometer showing the spatial damping (and growing) of a test wave at $f=\omega_1/2\pi=0.9$ MHz in a plasma with $\omega_{pi}/2\pi=1.3$ MHz. The top trace shows the case without the second wave. The middle trace is with the ω_2 wave at $\omega_2/2\pi=1.0$ MHz. Thus for this case $\omega_1<\omega_2$ and the ω_1 wave grows according to the theory. The bottom trace corresponds to the case with $\omega_2/2\pi=0.8$ MHz. Hence here $\omega_1>\omega_2$, and the test wave is transformed into the ω_2 wave.

4.4 Nonlinear Waves and Envelopes

4.4a Introduction

In this section, we introduce some exact nonlinear forms for waves and their envelopes which originate from typical nonlinear properties of the plasma. The approach we take here is an alternative to the Fourier harmonic approach we have used in previous sections. Instead of obtaining dynamics of the Fourier spectrum or the Fourier amplitude, we transform the original set of nonlinear equations that describe the wave dynamics to the lowest order into nonlinear equations whose properties are well known. This way, we can study the nonlinear properties of the wave development exactly in real space in the lowest order of the nonlinearity.

The nonlinear equations we use here are the Korteweg-deVries equation, the Burgers equation and the nonlinear Schrödinger equation. The first two equations describe the dynamics of the wave form itself, while the third describes the wave envelope.

In Subsection 4.4b, we introduce two stationary wave forms: "solitons" which are solutions of the Korteweg-deVries equation and which result from the balance between nonlinearity and *dispersion* and "shock" solutions of Burgers equation, which result from the balance between nonlinearity and *dissipation.* In Subsection 4.4c, we introduce stationary envelope forms (the envelope soliton, the envelope hole and the envelope shock) which result from the balance of nonlinearity and group velocity dispersion.

In Subsection 4.4d, we discuss the modulational instability, a nonlinear instability of a modulation envelope.

4.4b Nonlinear Waves – Solitons and Shocks

Let us consider the nonlinear term, $v\,\partial v/\partial x$, in the fluid equation of motion,

$$\frac{\partial v}{\partial t}+v\frac{\partial v}{\partial x}=F. \tag{4.160a}$$

From the Fourier analysis techniques, we have seen that such a term generates mode-mode coupling and higher temporal and spatial harmonics. In real space, this means a deformation of the wave form.

To see how the deformation of the wave takes place, let us linearize this expression around a constant velocity v_0,

$$\frac{\partial v_1}{\partial t}+v_0\frac{\partial v_1}{\partial x}=F. \tag{4.160b}$$

The solutions in the absence of F are easily seen to have the form $v_1=v_1(x-v_0t)$; for an initially sinusoidal perturbation of the form $\cos kx$, the solution is $\cos k(x-v_0t)$, that is, the phase velocity is given by v_0. Without linearization, we can expect the phase velocity to become proportional to $v_0+v_1\cos k(x-v_0t)$; this would mean that the top portion of the wave moves faster than the bottom portion and the wave steepens as shown in Fig. 45. This leads to an eventual overtaking of fluid elements and the breaking up of the wave, as shown in the right figure.

The presence of the self-consistent field F on the right hand side often produces an effect to prohibit such an overtaking, at least within

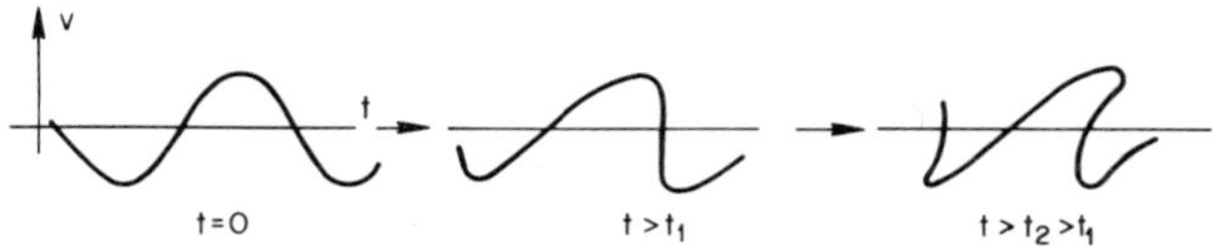

Fig. 45. Steepening and overtaking of a wave form due to the nonlinear phase velocity

some limited time scale. If we combine the force term with the field equation, F may be expressed as a function of v also. The lowest significant linear contribution of such a term will be $\partial^2 v/\partial t^2$ or $\partial^2 v/\partial x^2$. In Fourier analysis, such a term would correspond to a contribution to $\mathrm{Im}\,\omega$ or $\mathrm{Im}\,k$; hence, for a passive medium, these terms represent dissipation. If we take $\partial^2 v/\partial x^2$ as an example, the equation may be written as

$$\frac{\partial v}{\partial t}+v\frac{\partial v}{\partial x}-\alpha\frac{\partial^2 v}{\partial x^2}=0, \tag{4.161}$$

where α is a positive constant having a dimension of L^2/T. This equation is generally called Burgers equation. (The negative sign in front of $\alpha\,\partial^2 v/\partial x^2$ is introduced so that the solution is stable.) As the steepening progresses, the higher derivative term introduced above contributes more and when this term becomes comparable to the nonlinear term, the steepening is stopped.

We seek a stationary solution $v(\xi)$ of Eq. (4.161) in a moving frame $\xi=x-\lambda t$. If we change the variables t and x to ξ, Eq. (4.161) becomes

$$(v-\lambda)\frac{dv}{d\xi}=\frac{d^2 v}{d\xi^2}. \tag{4.162}$$

This can easily be integrated, using the condition that v is bounded as $\xi\to\pm\infty$, to give

$$v=\lambda\left[1-\tanh\left(\frac{\lambda\xi}{2\alpha}\right)\right]=\lambda\left[1-\tanh\frac{\lambda}{2\alpha}(x-\lambda t)\right]. \tag{4.163}$$

This expression represents a shock solution with the shock speed, the shock height and the shock thickness given by λ, λ and $\alpha\lambda^{-1}$ respectively.

The shock solution appears because of the introduction of the dissipative term, which increases entropy.

In the absence of dissipation, the lowest significant linear contribution of the force term will be $\partial^3 v/\partial x^3$. This term represents the lowest order *dispersion* effect. If we introduce this term into the Eq. (4.160) we have

$$\frac{\partial v}{\partial t}+v\frac{\partial v}{\partial x}+\beta\frac{\partial^3 v}{\partial x^3}=0, \tag{4.164}$$

where β is a constant having a dimension of L^3/T. This is called the Korteweg-deVries equation. One can verify easily that the third term contributes as a dispersive term rather than a dissipative term by applying Fourier transformation.

This dispersion term can also limit the wave steepening by a mechanism similar to that in the dissipative case.

As before, the stationary solution in the $\xi(=x-\lambda t)$ frame can readily be obtained. If we choose a boundary condition so that the solution is localized: $v\to 0$ and $dv/d\xi\to 0$ as $\xi\to\pm\infty$, the solution is

$$v=3\lambda\,\mathrm{sech}^2\sqrt{\frac{\lambda}{\beta}}\,\frac{\xi}{2}=3\lambda\,\mathrm{sech}^2\sqrt{\frac{\lambda}{\beta}}\left(\frac{x-\lambda t}{2}\right). \tag{4.165}$$

This represents a localized hump moving at a speed λ. Because of the remarkable stability of this solution against perturbations and collisions among different humps, such a wave form is often called a "soliton" (Zabusky and Kruskal, 1965. For the dynamic solution of the Korteweg-deVries equation, consult Gardner *et al.*, 1967).

As expected, the dispersion term does not increase entropy so that this solution is symmetric around $x=\lambda t$, an interesting contrast to the shock solution.

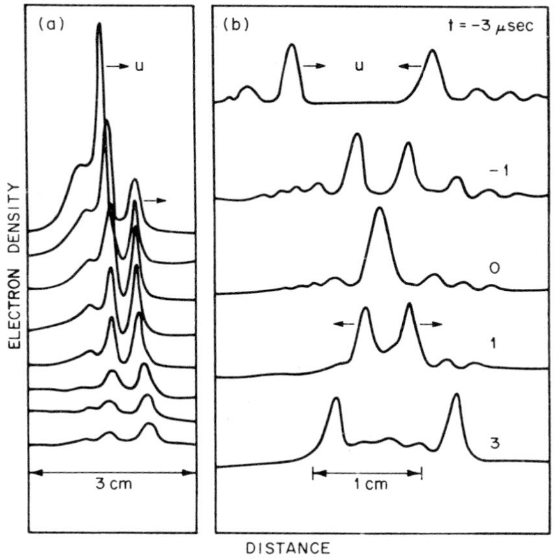

Fig. 46a and b. Interactions of two solitons. (a) Two solitons propagate in the same direction in the laboratory frame. The figure is depicted in the wave frame such that the smaller soliton is initially stationary. Time differences between adjacent curves are 10 μsec. (b) Two solitons propagating in directions opposite to each other, depicted in the laboratory frame. $\lambda_D\approx\approx 2\times 10^{-2}$ cm. (After Ikezi *et al.*, 1970)

We have seen that the wave steepening due to the nonlinear term in the equation of motion may be stopped in the presence of dissipation or dispersion and that a stationary shock or solition is formed. Consequently, if we can transform the original set of coupled nonlinear equations of motion and field equations which describe the desired wave dynamics into either Burgers equation or the Korteweg-deVries equation, we can study the dynamic development of the wave forms.

A general procedure for the transformation of a set of nonlinear partial differential equations into these equations has been obtained by Taniuti and Wei (1968). Using the same techniques, Washimi and Taniuti (1966) obtained solitary wave solution of the collisionless ion acoustic wave using the fluid description.

Experimental verification of ion acoustic solitons was made by Alikhanov *et al.* (1968) and Ikezi *et al.* (1970). In particular Ikezi *et al.*, have verified the stability of solitons against their mutual interactions (Fig. 46).

4.4c Nonlinear Envelopes – Envelope Solitons, Envelope Shocks, and Envelope Holes

Here we consider effects of nonlinearity and dispersion on the dynamics of the *envelope* of a plane wave. Unlike the case treated in Subsection 4.4b, we assume here a situation in which the wave has such strong dispersion that it does not allow either decay or harmonic generation. Namely we consider a case in which the carrier wave frequency ω_{k_0} satisfies neither the condition $\omega_{k_0}=\omega_{k_1}+\omega_{k_2}$ for $k_0=k_1+k_2$, nor the condition $2\omega_{k_0}\neq\omega_{2k_0}$ (for a one dimensional plane wave, k is a scalar and these requirements are relatively easily satisfied). We also assume that the wave particle interaction is negligible, which requires that $\omega_0/k_0 \gg v_T$ as well as $\partial\omega/\partial k|_{\omega_0} \gg v_T$ (v_T is the particle thermal speed). Under these conditions, the combined effects of nonlinearity and dispersion give rise to an interesting unique family of localized solutions for the wave envelope somewhat similar to the "solitons" that we have seen in the previous subsection.

Let us first construct the nonlinear equation for the *wave envelope* for such a case. We consider a modulated plane wave that can be expressed by $\mathrm{Re}\,\Phi(x,t)\,e^{i(k_0x-\omega_0t)}$. $\Phi(x,t)$ represents the slowly varying *complex* envelope function of the modulated wave with the carrier frequency and the wave number given by ω_0 and k_0 respectively.

When decay is prohibited, the lowest nonlinear effect originates from the interaction between the carrier wave and a virtual wave which is generated by the interaction between the ω_0 and $-\omega_0$ waves. For an electrostatic wave, this may be seen from Eq. (4.145), if we allow only the values $\pm k$. Thus,

$$\varepsilon_k^{(1)} \varphi_k + \left[3\varepsilon_k^{(3)}(k, k, -k) - \frac{4\varepsilon_k^{(2)}(k, 0)\varepsilon_{k-k}^{(2)}(k, -k)}{\varepsilon_{k-k}^{(1)}} \right] |\varphi_k|^2 \varphi_k = 0. \tag{4.166a}$$

Consequently, we can consider in such a case that the wave obeys a nonlinear dispersion relation given by

$$\varepsilon_{NL}(\omega, k, |\varphi_k|^2) = 0, \tag{4.166b}$$

or alternatively

$$\omega = \omega(k, |\varphi_k|^2). \tag{4.166c}$$

Hence we assume that the general plane wave we consider here also satisfies this type of nonlinear dispersion relation. (For an electromagnetic wave with more than one direction of polarization, modification of the simple theory here may be needed.) We write our nonlinear dispersion relation as

$$\omega = \omega(k, |\Phi|^2).$$

When the dispersion relation is given in this form, we can construct a wave equation for Φ in the following way. We expand ω around the carrier frequency ω_0 and the wave number k_0;

$$\omega - \omega_0 = \frac{\partial \omega}{\partial k_0}(k - k_0) + \frac{1}{2}\frac{\partial^2 \omega}{\partial k_0^2}(k - k_0)^2 + \frac{\partial \omega}{\partial |\Phi|_0^2}|\Phi|^2 \tag{4.167}$$

where $\partial\omega/\partial k_0$, $\partial^2\omega/\partial k_0^2$ and $\partial\omega/\partial|\Phi|_0$ are those values evaluated at $k = k_0$ and $\Phi = 0$. We took the lowest order term in the nonlinearity, but retained the second order term in the dispersion (group dispersion) because the $\partial\omega/\partial k_0$ term, as will be seen, simply represents the undistorted transmission of the wave packet at the group speed $\partial\omega/\partial k_0$. If we now substitute the operators $i\partial/\partial t$ for $(\omega - \omega_0)$ and $-i\partial/\partial x$ for $(k - k_0)$ into this expression and operate on Φ from the left, we obtain

$$i\left(\frac{\partial \Phi}{\partial t} + \frac{\partial \omega}{\partial k_0}\frac{\partial \Phi}{\partial x}\right) + \frac{1}{2}\frac{\partial^2 \omega}{\partial k_0^2}\frac{\partial^2 \Phi}{\partial x^2} - \frac{\partial \omega}{\partial |\Phi|_0^2}|\Phi|^2 \Phi = 0. \tag{4.168}$$

This is the desired wave equation for $\Phi(x, t)$. The first term represents the propagation of the packet at the group speed, the second term represents the linear distortion of the packet and the last term, the nonlinear distortion. Because of the resemblance to the Schrödinger equation, this equation is sometimes called the nonlinear Schrödinger equation.

As in the case of the Korteweg-deVries equation or Burgers equation, we now seek for a localized stationary solution of the nonlinear Schrödinger equation. For this purpose, we transform the variables x and t into ξ and τ where ξ is the coordinate moving at the group speed, i.e.,

$$\xi = x - (\partial\omega/\partial k_0)t \tag{4.169}$$

and τ is the normalized time given by

$$\tau = (\partial^2\omega/\partial k_0^2)t. \tag{4.170}$$

Thus Eq. (4.168) reduces to

$$i\frac{\partial\Phi}{\partial\tau} + \frac{1}{2}\frac{\partial^2\Phi}{\partial\xi^2} + \kappa|\Phi|^2\Phi = 0 \tag{4.171}$$

where

$$\kappa = -\frac{\partial\omega}{\partial|\Phi|_0^2}\bigg/\frac{\partial^2\omega}{\partial k_0^2}. \tag{4.172}$$

We note here that Φ is a complex function. Hence, we look for stationary solutions in $|\Phi|$, that is, a stationary shape of the packet. Because we are interested in a localized solution, we require the solution to be single-humped by imposing the following conditions:

1. $|\Phi|^2$ is bounded between two limits ϱ_S and ϱ_D;
2. at $|\Phi|^2 = \varrho_S$, $|\Phi|^2$ is an extremum, i.e., at $|\Phi|^2 = \varrho_S$, $\partial|\Phi|^2/\partial\xi = 0$ but $\partial^2|\Phi|^2/\partial\xi^2 \neq 0$;
3. ϱ_D is the asymptotic value of $|\Phi|^2$ as $\xi \to \pm\infty$, i.e., at $|\Phi|^2 = \varrho_D$, $\partial^n|\Phi|^2/\partial\xi^n = 0$, $n = 1, 2, \ldots$.

We now look for a solution of Eq. (4.171) which satisfies these conditions. For this we introduce two real variables ϱ and σ which represent the real and the imaginary parts of Φ:

$$\Phi(\xi,\tau) = \sqrt{\varrho(\xi,\tau)}\, e^{i\sigma(\xi,\tau)}. \tag{4.173}$$

Substituting this expression into Eq. (4.171), we have

$$\frac{\partial\varrho}{\partial\tau} + \frac{\partial}{\partial\xi}\left(\varrho\frac{\partial\sigma}{\partial\xi}\right) = 0 \tag{4.174}$$

and

$$\frac{1}{8}\frac{d}{d\varrho}\left[4\kappa\,\varrho^2+\frac{1}{\varrho}\left(\frac{\partial\varrho}{\partial\xi}\right)^2\right]=\frac{\partial\sigma}{\partial\tau}+\frac{1}{2}\left(\frac{\partial\sigma}{\partial\xi}\right)^2. \tag{4.175}$$

The stationary condition for $|\Phi|^2$ $(=\varrho)$ gives $\partial\varrho/\partial\tau=0$. Hence from Eq. (4.174), we have

$$\varrho\,\frac{\partial\sigma}{\partial\xi}=c(\tau). \tag{4.176a}$$

We show now that the only choice for the integration constant $c(\tau)$ is a constant independent of τ. To prove this we note that the left hand side of Eq. (4.175) is a function of ξ only, thus

$$\frac{\partial\sigma}{\partial\tau}+\frac{1}{2}\left(\frac{\partial\sigma}{\partial\xi}\right)^2=f(\xi).$$

By taking derivatives with respect to τ and ξ, we have

$$\frac{\partial^3\sigma}{\partial\tau^2\,\partial\xi}-\frac{1}{\varrho^3}\,\frac{d\varrho}{d\xi}\,\frac{dc^2}{d\tau}=0,$$

while from Eq. (4.176a),

$$\frac{\partial^3\sigma}{\partial\tau^2\,\partial\xi}=\frac{1}{\varrho}\,\frac{d^2c}{d\tau^2}.$$

Hence

$$\frac{1}{\varrho}\,\frac{d^2c}{d\tau^2}-\frac{1}{\varrho^3}\,\frac{d\varrho}{d\xi}\,\frac{dc^2}{d\tau}=0$$

or

$$\frac{d^2c}{d\tau^2}\Big/\frac{dc^2}{d\tau}=\frac{1}{\varrho^2}\,\frac{d\varrho}{d\xi}=\text{const.}$$

Because we cannot accept the solution $\varrho^{-2}\,d\varrho/d\xi=\text{const}$, the only alternative choice is $c(\tau)=\text{const}$. Consequently Eq. (4.176a) becomes

$$\varrho\,\frac{\partial\sigma}{\partial\xi}=c_1\,(\text{const}) \tag{4.176b}$$

or

$$\sigma=\int\frac{c_1}{\varrho}\,d\xi+A(\tau). \tag{4.177a}$$

Because $\partial\sigma/\partial\xi$ is proved to be a function of ξ only, from Eq. (4.175), $\partial\sigma/\partial\tau$ should also be a function of ξ. Hence we take $dA/d\tau$ to be constant $(=\Omega)$,

$$\sigma=\int\frac{c_1}{\varrho}d\xi+\Omega\tau. \tag{4.177b}$$

If we use this expression in Eq. (4.175), we obtain the following ordinary differential equation for $\varrho(\xi)$,

$$\left(\frac{d\varrho}{d\xi}\right)^2=-4\kappa\varrho^3+8\Omega\varrho^2+c_2\varrho-4c_1^2. \tag{4.178}$$

We now seek a solution of this equation subject to the imposed conditions 1 to 3. At this stage it is convenient to divide the problem into two cases $\kappa>0$ and $\kappa<0$.

Case A: $\kappa(=-\partial\omega/\partial|\Phi|_0^2/\partial^2\omega/\partial k_0^2)>0$.

To satisfy condition 1, $d\varrho/d\xi$ should vanish only at two values of ϱ, ϱ_D and ϱ_S. In addition, for the root at ϱ_D to represent an asymptotic value of ϱ, it should be a double root. When $\kappa>0$, these conditions are met only for $-4c_1^2\geq 0$, or $c_1=0$, hence also $c_2=0$. Eq. (4.178) then reduces to

$$\left(\frac{d\varrho}{d\xi}\right)^2=-4\kappa\varrho^3+8\Omega\varrho^2=-4\kappa\varrho^2(\varrho-\varrho_S) \tag{4.179}$$

where

$$\varrho_S=2\Omega/\kappa. \tag{4.180}$$

Eq. (4.179) can then be integrated easily to give

$$\varrho=\varrho_0\,\mathrm{sech}^2(\sqrt{\kappa\varrho_0}\,\xi) \tag{4.181a}$$

where $\varrho_0(=\varrho_S)=2\Omega/\kappa$, $\Omega>0$ and

$$\sigma=\Omega\tau. \tag{4.182a}$$

This solution has a shape similar to the soliton solution for a wave in a weakly dispersive nonlinear medium (Subsection 4.4b). Hence this may be called an envelope soliton. A distinctive difference, however, is that the envelope soliton introduced here moves at a speed $\partial\omega/\partial k_0$ which is independent of the height of the soliton. In fact, because the nonlinear Schrödinger Eq. (4.171) can be shown to be satisfied by another function $\Phi'(\xi,\tau)$ given by

$$\Phi'(\xi,\tau)=e^{i(u\xi-\frac{1}{2}u^2\tau)}\,\Phi(\xi-u\tau,\tau)$$

(the Galilean transformation), the speed of the soliton can be chosen to be independent of $\partial\omega/\partial k_0$ as well as of ϱ_0. With the additional independent parameter u, the solitary envelope solution becomes

$$\varrho' = \varrho_0 \operatorname{sech}^2\left[\sqrt{\kappa\varrho_0}(\xi - u\tau)\right], \tag{4.181b}$$

$$\sigma' = u\xi + (\Omega - \tfrac{1}{2}u^2)\tau. \tag{4.182b}$$

Case B: $\kappa(= -\partial\omega/\partial|\Phi|_0^2/\partial^2\omega/\partial k_0^2) < 0$.

When $\kappa < 0$, we see from Eq. (4.178) that the double root appears for a larger value of ϱ than for the single root. Consequently we can accept here a finite value of c_1^2. Eq. (4.178) can then be expressed

$$\left(\frac{d\varrho}{d\xi}\right)^2 = 4|\kappa|(\varrho - \varrho_1)^2(\varrho - \varrho_S). \tag{4.183}$$

Integrating this equation we obtain the solution

$$\varrho = \varrho_1\left[1 - a^2 \operatorname{sech}^2\left(\sqrt{|\kappa|\varrho_1}\, a\xi\right)\right], \tag{4.184a}$$

where

$$a^2 = \frac{\varrho_1 - \varrho_S}{\varrho_1} \leq 1 \tag{4.185}$$

and

$$c_1^2 = |\kappa|\varrho_1^3(1 - a^2), \tag{4.186}$$

$$|\kappa|\varrho_1(3 - a^2) = -2\Omega \tag{4.187}$$

with

$$\begin{aligned}\sigma &= \int \frac{c_1}{\varrho} d\xi + \Omega\tau \\ &= \sin^{-1}\left\{a \tanh\left(\sqrt{|\kappa|\varrho_1}\, a\xi\right) / \left[1 - a^2 \operatorname{sech}^2\left(\sqrt{|\kappa|\varrho_1}\, a\xi\right)\right]^{1/2}\right\} + \Omega\tau.\end{aligned} \tag{4.188a}$$

The above solution represents an inversed envelope soliton, i.e., a soliton for the *absent* region of the wave intensity $|\Phi|^2$. Hence this may be called an envelope hole. Compared with the envelope soliton, the envelope hole has an additional independent parameter "a" which designates the depth of modulation. When $a = 1$, the extremum reaches $\varrho = 0$ and the solution can be expressed as

$$|\Phi| = \pm\sqrt{\rho_1} \tanh\left(\sqrt{\kappa\rho_1}\,\xi\right). \tag{4.189}$$

Consequently this solution may be called an envelope shock. The envelope hole solution can also be generalized by the Galilean transformation to give

$$\varrho' = \varrho_1 \{1 - a^2 \operatorname{sech}^2 [\sqrt{|\kappa| \varrho_1}\, a(\xi - u\tau)]\}, \tag{4.184b}$$

$$\sigma' = \sigma + u\xi - \tfrac{1}{2} u^2 \tau. \tag{4.188b}$$

Typical shapes of envelope solitons and holes are plotted in Fig. 47.

As in the case of the soliton, these solutions for the envelope function can be shown to be stable against perturbations and collisions (Hasegawa and Tappert, 1973a, b) and can be regarded as representing asymptotic envelope wave forms in a nonlinear dispersive medium (Whitham, 1965a, b; Lighthill, 1965; Karpman and Krushkal, 1968; Zakharov and Shabat, 1971).

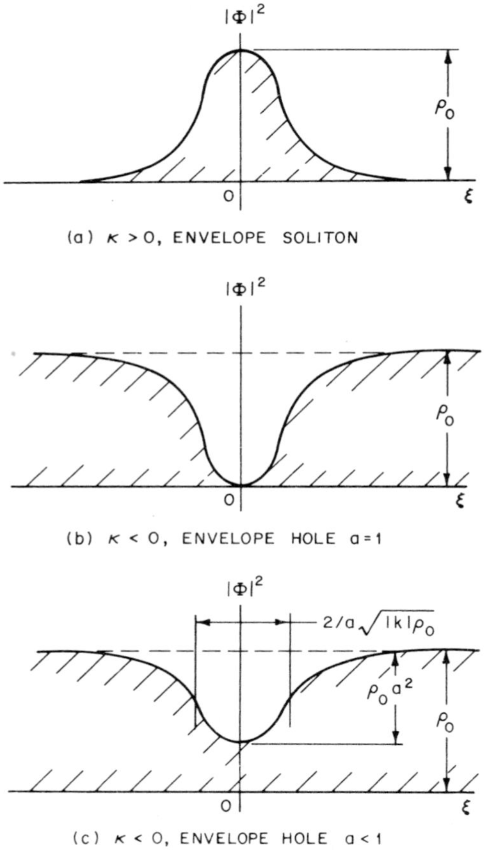

Fig. 47a–c. Localized stationary solutions for the nonlinear Schrödinger equation (4.171) for $\kappa > 0$ (envelope soliton) and for $\kappa > 0$ (envelope hole)

4.4d Modulational Instability

Let us now study the dynamics of a continuous train of modulated waves in a nonlinear and dispersive medium of the type discussed in the previous subsection, a medium which prohibits neither decay nor harmonic generation. In this case we must assume a finite value for Φ as $\xi \to \pm\infty$, thus the suitable nonlinear Schrödinger equation must contain an asymptotic value of $\Phi(=\Phi_0)$, and from Eq. (4.171) we have

$$i\frac{\partial \Phi}{\partial \tau}+\frac{1}{2}\frac{\partial^2 \Phi}{\partial \xi^2}+\kappa(|\Phi|^2-|\Phi_0|^2)\Phi=0. \tag{4.190}$$

To study the stability of the modulation envelope described by this equation, we again introduce the variables ϱ and σ defined by

$$\Phi=\sqrt{\varrho(\xi,\tau)}\,e^{i\sigma(\xi,\tau)}. \tag{4.173}$$

If we substitute this expression into Eq. (4.190), we obtain

$$\varrho_\tau+(\varrho\sigma_\xi)_\xi=0 \tag{4.174}$$

and

$$\kappa(\varrho-\varrho_0)+\frac{\varrho_{\xi\xi}}{4\varrho}-\sigma_\tau-\frac{1}{2}\sigma_{\xi\xi}^2-\frac{1}{8}\frac{\varrho_\xi^2}{\varrho^2}=0 \tag{4.191}$$

where the subscripts τ and ξ indicate partial derivatives with respect to these variables. Here, instead of a stationary solution, we are interested in the dynamic solution. Hence we linearize the above set of equations for ϱ and σ in the form,

$$\begin{pmatrix}\varrho\\ \sigma\end{pmatrix}=\begin{pmatrix}\varrho_0\\ 0\end{pmatrix}+\mathrm{Re}\begin{pmatrix}\varrho_1\\ \sigma_1\end{pmatrix}e^{i(K\xi-\Omega\tau)}. \tag{4.192}$$

If we substitute this expression into Eqs. (4.174) and (4.191), we obtain the following dispersion relation for Ω and K:

$$\Omega^2=K^4/4-\kappa\varrho_0K^2=\frac{1}{4}(K^2-2\kappa\varrho_0)^2-\kappa^2\varrho_0^2. \tag{4.193}$$

From this dispersion relation, we can see immediately that if $\kappa>0$, Ω^2 becomes negative for a small value of K. That is, for a long wavelength perturbation, the modulation becomes unstable and grows. The maximum growth is attained for $K=\sqrt{2\kappa\varrho_0}$ and the corresponding growth rate is given by $\kappa\varrho_0=\kappa|\Phi_0|^2$. The critical wave number K_c for stability is given by

$$K_c = 2\sqrt{\kappa \varrho_0}, \tag{4.194}$$

and a perturbation whose wavelength λ is longer than $2\pi/K_c$ becomes unstable. It is interesting to note that the argument of the stationary solitons we obtained in Subsection 4.4b is just one half of K_c.

The instability derived above is called the modulational instability. As found, the modulational instability occurs in a medium in which the nonlinear frequency shift $\partial\omega/\partial|\Phi|_0^2$ and the group dispersion $\partial^2\omega/\partial k^2$ have opposite signs. The physical mechanism of this nonlinear instability may be interpreted as follows. When $\kappa>0$, we can regard Eq. (4.190) as the Schrödinger equation for quasiparticles whose wave function is given by Φ, and that are trapped by a self-generated potential $V=\kappa(|\Phi|^2-|\Phi_0|^2)$. If $\kappa>0$, this potential has an attractive sign.

If we increase the quasiparticle density $|\Phi|^2$, the potential depth increases; thus if $\kappa>0$, more quasiparticles are attracted, leading to a further increase in the potential depth. In this respect the instability may be regarded as a consequence of the self-trapping of the quasiparticles.

Many waves in a plasma satisfy the condition of the modulational instability, for example, electron and ion cyclotron waves (Taniuti and Washimi, 1968; Tam, 1969; Hasegawa, 1970, 1971). If we write one transverse component of the perturbed magnetic field $B_1=\Phi B_0$, the nonlinear Schrödinger equation for a cyclotron wave propagating along the magnetic field becomes (Hasegawa, 1971),

$$i\frac{\partial\Phi}{\partial\tau}+\frac{1}{2}\frac{\partial^2\omega}{\partial k_0^2}\frac{\partial^2\Phi}{\partial\xi^2}-\frac{1}{4}\frac{k v_A^2}{v_g}(|\Phi|^2-|\Phi_0|^2)\Phi=0, \tag{4.195}$$

where v_A is the Alfvén speed and v_g is the group speed. This expression can apply to either electron or ion cyclotron waves if appropriate values for v_g and $\partial^2\omega/\partial k_0^2$ are used. For both waves, $v_g>0$ for the entire propagation regime of the waves, hence the unstable frequency range is decided simply by the condition $\partial^2\omega/\partial k_0^2<0$. For the ion cyclotron wave, $\partial^2\omega/\partial k^2$ is always negative for $0<\omega<\omega_{ci}$, hence it is always unstable; for the electron cyclotron wave $\partial^2\omega/\partial k^2<0$ for $\omega_{ce}/4<\omega<\omega_{ce}$, hence it is unstable only in this frequency range.

Fig. 48 shows the results of computer experiments on the modulational instability of the ion cyclotron wave (Hasegawa, 1971). The initial amplitudes of the waves are both $B_1/B_0=0.123$. The difference between the left and the right plot exists simply in the initial depth of modulation. The theoretical growth rate is $0.023\,\omega_c$ and agrees well with the experimental results. Because the cyclotron wave is circularly polarized, its intensity $|B_y^2+B_z^2|^{1/2}$ represents the physical envelope of the

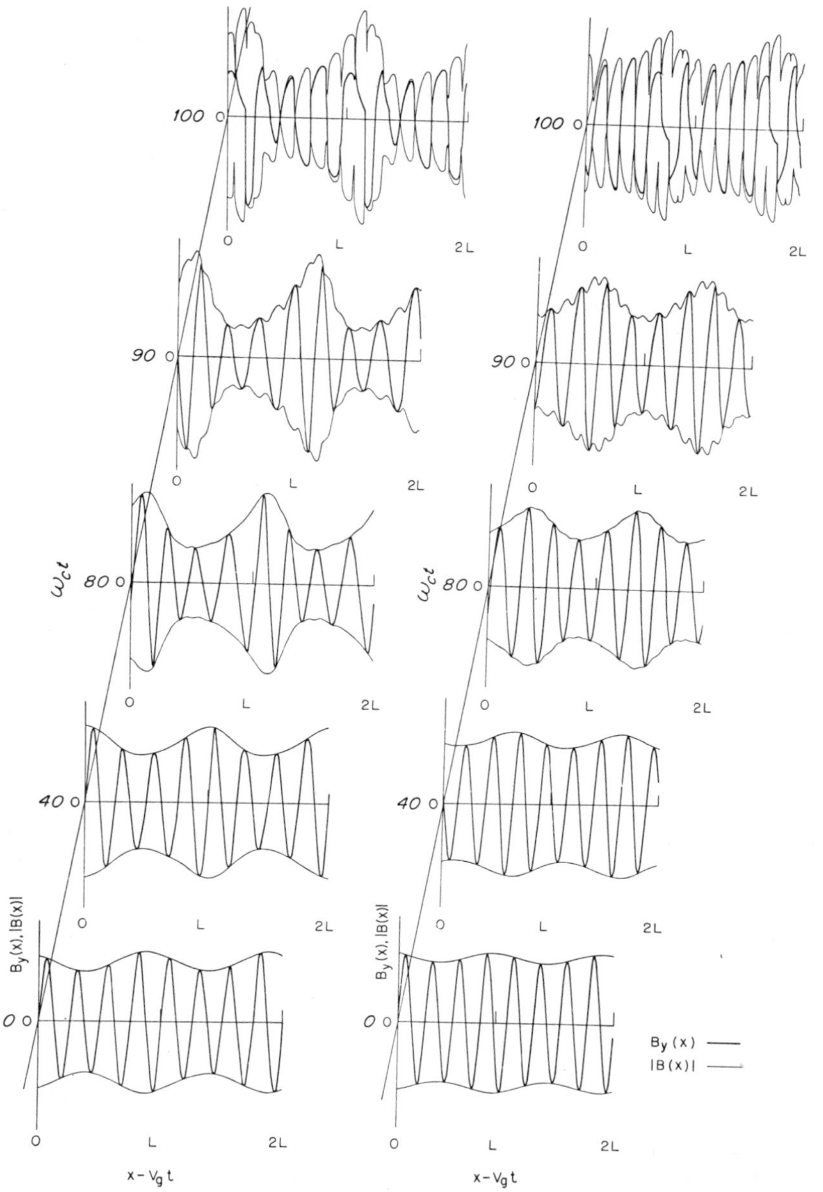

Fig. 48. Results of computer experiments on the modulational instability of the ion cyclotron wave (Hasegawa, 1971). Plotted are the spatial variation of one transverse component of the wave magnetic field B_y (solid curve) and the magnitude of the wave intensity $|B|$ (thin curve) at different times of evolution. The initial amplitude $|B|/B_0$ is 0.123 for both cases. The left has an initially larger depth of modulation than the right

wave $|\Phi|$ as long as the wave remains circularly polarized. The breaking up of the wave seen at the last time step is a consequence of the particle trajectory crossing in the direction of propagation. The particles are accelerated by the intensity gradient of the wave magnetic field which is increased rapidly by the instability.

In reality, the modulational instability is a rather weak instability because the characteristic growth rate is (amplitude)^2 × (characteristic frequency). In particular, for an electron mode, such as the electron cyclotron wave, electrons must drag ions for the waves to be self-trapped, hence the linear characteristic frequency in the above expression is the ion cyclotron frequency rather than the electron cyclotron frequency. Consequently, the modulational instability may not be observable for whistler waves in the magnetosphere.

On the other hand, whistlers observed in the upstream direction of the solar wind have in general large amplitudes ($B_1/B_0 \sim 0.3$) and thus may well be subject to the modulational instability. Wave packets observed by Russell *et al.* (1971) as was shown in Fig. 10 seem to support the evidence of the modulational instability.

References for Chapter 4

Alikhanov, S.G., Belan, V.G., Sagdeev, R.Z.: Nonlinear ion acoustic waves in a plasma. Zh. Eksp. Teor. Fiz. Pis'ma **7**, 405 (1968)

Davidson, R.C.: Methods in nonlinear plasma theory. New York-London: Academic Press, 1972

Davidson, R.C., Kaufman, A.N.: On the kinetic equation for resonant three-wave coupling. J. Plasma Phys. **3**, 97 (1969)

Drummond, W.E., Malmberg, J.H., O'Neil, T.M., Thompson, J.R.: Nonlinear development of the beam-plasma instability. Phys. Fluids **13**, 2422 (1970)

Drummond, W.E., Pines, D.: Nonlinear stability of plasma oscillations. Nucl. Fusion Suppl. Pt. **3**, 1049 (1962)

Dupree, T.H.: A perturbation theory for strong plasma turbulence. Phys. Fluids **9**, 1773 (1966)

Gardner, C.S., Green, J.M., Kruskal, M.D., Miura, R.M.: Method for solving the Korteweg-deVries equation. Phys. Rev. Letters **19**, 1095 (1967)

Gentle, K.W., Lohr, J.: Experimental determination of the nonlinear interaction in a one dimensional beam-plasma system. Phys. Fluids **16**, 1464 (1973)

Hasegawa, A.: Observation of self-trapping instability of a plasma cyclotron wave in a computer experiment. Phys. Rev. Letters **24**, 1165 (1970)

Hasegawa, A.: Theory and computer experiment on self-trapping instability of plasma cyclotron waves, Phys. Fluids **15**, 870 (1971)

Hasegawa, A., Tappert, F.D.: Transmission of stationary nonlinear optical pulses in dispersive dielectric fibers. I. Anomalous dispersion. Appl. Phys. Letters **23**, 142 (1973a)

Hasegawa, A., Tappert, F.D.: Transmission of stationary nonlinear optical pulses in dispersive dielectric fibers. II. Normal dispersion, Appl. Phys. Letters **23**, 171 (1973b)

Ikezi, H., Kiwamoto, Y.: Observations of nonlinear Landau damping of ion acoustic waves. Phys. Rev. Letters **27**, 718 (1971)

Ikezi, H., Taylor, R.J., Baker, D.R.: Formation and interaction of ion acoustic solitons. Phys. Rev. Letters **25**, 11 (1970)

Karpman, V.I., Krushkal, E.M.: Modulated waves in nonlinear dispersive media. Zh. Eksp. Teor. Fiz. **55**, 530 (1968) [Soviet Phys. JETP English Transl. **28**, 277 (1969)]

Kono, M., Ichikawa, Y.H.: Renormalization of the wave-particle interaction in weakly turbulent plasmas. Progr. Theor. Phys. **49**, 754 (1973)

Lamb, G.L., Jr.: Analytical descriptions of ultrashort optical pulse propagation in a resonant medium. Rev. Modern Phys. **43**, 99 (1971)

Lighthill, M.J.: Contribution to the theory of waves in nonlinear dispersive system. J. Inst. Math. Appl. **1**, 269 (1965)

Mima, K.: Modification of weak turbulence theory due to perturbed orbit effects, I. General formulation, J. Phys. Soc. Japan **34**, 1620 (1973)

Nishikawa, K.: Interaction between an electron wave and an ion wave due to scattering by electrons. J. Phys. Soc. Japan **29**, 449 (1970)

Nishikawa, K., Wu, C.-S.: Effect of electron trapping on the ion wave instability. Phys. Rev. Letters **23**, 1020 (1969)

Nozaki, K., Taniuti, T.: Propagation of solitary pulses in interactions of plasma waves. J. Phys. Soc. Japan **34**, 796 (1973)

O'Neil, T.M., Winfrey, J.H.: Nonlinear interaction of a small cold beam and a plasma, part II. Phys. Fluids **15**, 1514 (1972)

O'Neil, T.M., Winfrey, J.H., Malmberg, J.H.: Nonlinear interaction of a small cold beam and a plasma. Phys. Fluids **14**, 1204 (1971)

Onishchenko, I.N., Linetskii, A.R., Matsiborko, N.G., Shapiro, V.D., Shevchenko, V.I.: Contribution to the nonlinear theory of excitation of a monochromatic plasma wave by an electron beam, Zh. Eksp. Teor. Fiz. Pis. red. **12**, 407 (1970)

Oya, H.: Verification of theory of weak turbulence relating to the sequence of diffuse plasma resonances in space. Phys. Fluids **14**, 2487 (1971)

Roberson, C., Gentle, K.W.: Experimental test of the quasilinear theory of the gentle bump instability. Phys. Fluids **14**, 2462 (1971)

Rogister, A.: Nonlinear estabilization of $\boldsymbol{E} \times \boldsymbol{B}$ electron drift instability with $T_i \sim T_e$. Phys. Rev. Letters **30**, 86 (1973)

Romanov, U.A., Filippov, G.: The interaction of fast electron beams with longitudinal plasma waves. Zh. Eksp. Teor. Fiz. **40**, 123 (1961) [Soviet Phys. JETP English Transl. **13**, 87 (1961)]

Rudakov, L.I., Tsytovich, V.N.: The theory of plasma turbulence for strong wave-particle interaction. Plasma Phys. **13**, 213 (1971)

Russell, C.T., Childers, D.D., Coleman, P.J., Jr.: OGO5 observations of upstream waves in the interplanetary medium: discrete wave packets. J. Geophys. Res. **76**, 845 (1971)

Sagdeev, R.Z., Galeev, A.A.: Nonlinear plasma theory, ed. by T.M. O'Neil and D.L. Book. New York: W.A. Benjamin, Inc. 1969

Sato, T.: Nonlinear theory of cross-field instability-explosive mode coupling. Phys. Fluids **14**, 2426 (1971)

Tam, C.K.W.: Amplitude dispersion and nonlinear instability of whistlers. Phys. Fluids **12**, 1028 (1969)

Taniuti, T.: Nonlinear space charge effects in cold plasma. J. Phys. Soc. Japan **18**, 408 (1963)

Taniuti, T., Washimi, H.: Self-trapping and instability of hydromagnetic waves along the magnetic field in a cold plasma. Phys. Rev. Letters **21**, 209 (1968)

Taniuti, T., Wei, C.-C.: Reductive perturbation method in nonlinear wave propagation. J. Phys. Soc. Japan **24**, 941 (1968)

Vedenov, A. A., Velikhov, E. P., Sagdeev, R. Z.: Nonlinear oscillations of a rarefield plasma. Nucl. Fusion **1**, 82 (1961)
Washimi, H., Taniuti, T.: Propagation of ion-acoustic waves of small amplitude. Phys. Rev. Letters **17**, 996 (1966)
Weinstock, J.: Formation of statistical theory of strong plasma turbulence. Phys. Fluids **12**, 1045 (1969)
White, R. B., Lee, Y. C., Nishikawa, K.: Nonlinear mode coupling and relaxation oscillations. Phys. Rev. Letters **29**, 1315 (1972)
Whitham, G. B.: Nonlinear dispersive waves. Proc. Roy. Soc. (London) **A 283**, 238 (1965a)
Whitham, G. B.: A general approach to linear and nonlinear dispersive waves using a Lagrangian, J. Fluid Mech. **22**, 273 (1965b)
Zabusky, N. J., Kruskal, M. D.: Interaction of "solitons" in a collisionless plasma and the reoccurrence of initial states. Phys. Rev. Letters **15**, 240 (1965)
Zakharov, V. E., Shabat, A. B.: Exact theory of two-dimensional self-focusing and one-dimensional self-modulation of waves in nonlinear media. Zh. Eksp. Teor. Fiz. **61**, 118 (1971) [Soviet. Phys. JETP English Transl. **34**, 62 (1972)]
Zaslavskii, G. M., Sagdeev, R. Z.: Limit of statistical description of a nonlinear wave field. Zh. Eksp. Teor. Fiz. **52**, 1081 (1967) [Soviet. Phys. JETP English Transl. **25**, 718 (1967)]

Appendix

Summary of Necessary Conditions of Plasma Instabilities

1. Microinstabilities

1.1 Two humped velocity distribution parallel to the magnetic field

a) $\boldsymbol{k}\times\boldsymbol{E}_1=0$, $\boldsymbol{k}\|\boldsymbol{B}_0$ ($\boldsymbol{k}\nparallel\boldsymbol{B}_0$ can also be unstable)

Necessary conditions:

$$\frac{\partial f_0}{\partial v}>0 \quad \text{or} \quad v_0>v_{Te}, \qquad \omega\sim\omega_{pe}$$

$$v_0>c_s, \qquad \omega\lesssim\omega_{pi}$$

b) $\boldsymbol{k}\cdot\boldsymbol{E}_1=0$, $\boldsymbol{k}\|\boldsymbol{B}_0$ ($\boldsymbol{k}\nparallel\boldsymbol{B}_0$ can also be unstable)

Necessary condition:

$$v_0>v_A, \qquad \omega\lesssim\omega_c$$

c) in the magnetic mirror, $\boldsymbol{k}\|\boldsymbol{B}_0$

Necessary condition:

$$\frac{\partial f}{\partial w}>0, \qquad \omega\sim n\omega_b$$

1.2 Anisotropic velocity distributions

a) $\boldsymbol{k}\times\boldsymbol{E}_1=0$, $\boldsymbol{k}\nparallel\boldsymbol{B}_0$, $\boldsymbol{k}\not\perp\boldsymbol{B}_0$

Necessary condition:

$$T_\perp/T_\| >2\mathrm{n}, \quad \mathrm{n}=1,2,\ldots, \omega\sim\omega_c(n+1/2)$$

b) $\boldsymbol{k}\times\boldsymbol{E}_1=0$, $\boldsymbol{k}$ almost $\perp\boldsymbol{B}_0$, or $\boldsymbol{k}\perp\boldsymbol{B}_0$

Necessary condition:

$$\int_0^\infty\left[J_{\mathrm{n}}\left(\frac{k_\perp v_\perp}{\omega_c}\right)\right]^2\frac{1}{v_\perp}\frac{\partial f_0}{\partial v_\perp}2\pi v_\perp dv_\perp>0,$$

$$\omega\sim\mathrm{n}\omega_c, \qquad \mathrm{n}=0,1,2,\ldots$$

c) $\boldsymbol{k}\cdot\boldsymbol{E}_1=0$, $\boldsymbol{k}\parallel\boldsymbol{B}_0$ ($\boldsymbol{k}\nparallel\boldsymbol{B}_0$ can also be unstable)
Necessary condition:

$$\frac{T_\perp}{T_\parallel}>\frac{\omega_c}{\omega_c-\omega},\qquad \omega\lesssim\omega_c$$

cold plasma concentration

d) in the presence of nonuniformity perpendicular to the magnetic field, $\boldsymbol{k}\times\boldsymbol{E}_1=0$, $\boldsymbol{k}\perp\boldsymbol{B}_0$
Necessary condition:
loss cone distribution,

$$\omega\sim \mathrm{n}\omega_{ci},\qquad \mathrm{n}=1,2,\ldots$$

1.3 Hydromagnetic instabilities due to anisotropic pressures

$$\boldsymbol{k}\cdot\boldsymbol{E}_1\neq 0,\qquad \boldsymbol{k}\times\boldsymbol{E}_1\neq 0,$$

Hose instability:

$\boldsymbol{k}$ almost $\parallel\boldsymbol{B}_0$ or $\boldsymbol{k}\parallel\boldsymbol{B}_0$

Necessary condition:

$$1-\sum_{\text{species}}\frac{1}{2}(\beta_\parallel-\beta_\perp)<0,\qquad \omega\sim 0$$

Mirror instability:

$\boldsymbol{k}$ almost $\perp\boldsymbol{B}_0$

Necessary condition:

$$1+\sum_{\text{species}}\beta_\perp\left(1-\frac{\beta_\perp}{\beta_\parallel}\right)<0,\qquad \omega\sim 0$$

1.4 Instabilities in partially ionized plasmas

$\boldsymbol{k}\times\boldsymbol{E}_1=0$, $\boldsymbol{k}$ arbitrary direction

Necessary conditions:

if $\omega_{ci}\tau_i\ll 1$, $\boldsymbol{E}_0\cdot\nabla n_0>0$, $\omega\sim 0$

if $\omega_{ci}\tau_i\gg 1$, $\omega_e^*>kc_s$ or $v_E>c_s$, $\omega\sim kc_s$

II. Macroinstabilities

2.1 Drift wave instabilities

a) low to medium β,

$\boldsymbol{k}\times\boldsymbol{E}_1\sim 0$, $\boldsymbol{k}$ almost $\perp\boldsymbol{B}_0$

Necessary conditions:

$v_{Te}>\omega/k_\parallel>v_{Ti}$, $\beta\leqq 0.13$

$k_\perp\varrho_i\neq 0$ or $(\nabla T)\cdot(\nabla n_0)<0$, $\omega\sim\omega_e^*$

cold electron stabilization

b) high β,

$$\boldsymbol{k}\times\boldsymbol{E}_1 \neq 0, \quad \boldsymbol{k}\cdot\boldsymbol{E}_1 \neq 0, \quad \boldsymbol{k} \nparallel \boldsymbol{B}_0$$

Necessary conditions:

$$v_D > v_A, \quad \text{or} \quad (\nabla T)\cdot(\nabla n_0) < 0,$$

$$\omega \sim k v_A$$

cold plasma destabilization

2.2 Rayleigh-Taylor instability,

$$\boldsymbol{k}\times\boldsymbol{E}_1 = 0, \quad \boldsymbol{k}\perp\boldsymbol{B}_0$$

Necessary condition:

$$\boldsymbol{g}\cdot\nabla n_0 < 0, \quad \omega \sim 0$$

cold electron stabilization if $\boldsymbol{k} \not\perp \boldsymbol{B}_0$, finite Larmor radius stabilization

2.3 Kelvin-Helmholtz instabilities

a) hydromagnetic

$$\boldsymbol{k}\cdot\boldsymbol{v}_1 = 0, \quad \boldsymbol{k}\perp\boldsymbol{B}_0 \quad \text{or} \quad \boldsymbol{k} \text{ almost } \perp\boldsymbol{B}_0$$

Necessary condition:

$$(\boldsymbol{k}\cdot\boldsymbol{v}_0)^2 > \left(\frac{1}{n_{0\mathrm{I}}}+\frac{1}{n_{0\mathrm{II}}}\right)[n_{0\mathrm{I}}(\boldsymbol{k}\cdot\boldsymbol{v}_{A\mathrm{I}})^2 + n_{0\mathrm{II}}(\boldsymbol{k}\cdot\boldsymbol{v}_{A\mathrm{II}})^2], \ \omega \sim \boldsymbol{k}\cdot\boldsymbol{v}_0$$

$\boldsymbol{k}\cdot\boldsymbol{v}_1 \neq 0$ is also possible.

b) electrostatic

$$\boldsymbol{k}\times\boldsymbol{E}_1 = 0, \quad \boldsymbol{k}\perp\boldsymbol{B}_0$$

Necessary condition:

$$2ka \leqq 1.3, \quad \omega \sim 0$$

2.4 Current Pinch

a) cylindrical pinch

sausage: $B_{0\Theta}^2 > 2B_{0z}^2$
kink: $B_{0\Theta}^2 \ln(L/a) > B_{0z}^2$
helical: $B_{0\Theta} > (2\pi a/L) B_{0z}$

b) sheet current

$$\boldsymbol{k}\cdot\boldsymbol{v}_1 = 0, \quad \boldsymbol{k}\perp\boldsymbol{J}_0$$

Necessary condition:
finite resistivity plus nonuniform current

General References

I. Plasma Physics

Allis, W. P., Buchsbaum, S. J., Bers, A.: Waves in anisotropic plasmas. Cambridge, Massachusetts; M. I. T. Press, 1963

Bekefi, G.: Radiation processes in plasmas. New York: Wiley, 1966

Davidson, R. C.: Methods in nonlinear plasma theory. New York: Academic Press, 1972

Jackson, J. D.: Classical electrodynamics. New York: Wiley, 1962

Kadomtsev, B. B.: Plasma turbulence. New York: Academic Press, 1965

Klimontovich, Y. L.: The statistical theory of non-equilibrium processes in a plasma. (transl. by H. S. H. Massey and O. M. Blunn, ed. by D. Ter Haar). Cambridge, Massachusetts: M. I. T. Press, 1967

Landau, L. D., Lifshitz, E. M.: Electrodynamics of continuous media. Reading, Massachusetts: Addison-Wesley, 1959

Leontovich, M. A. (ed.): Reviews of plasma physics (vols. 1–5) (transl. by H. Lashinsky). New York: Consultants Bureau, 1965–1968

Longmire, C. L.: Elementary plasma physics. New York: Interscience Publishers, 1963

Rose, D. J., Clark, M., Jr.: Plasmas and controlled fusion. Cambridge, Massachusetts: M. I. T. Press, 1961

Sagdeev, R. Z., Galeev, A. A.: Nonlinear plasma theory (revised and ed. by T. M. O'Neil and D. L. Book). New York: Benjamin, 1969

Stix, T. H.: The theory of plasma waves. New York: McGraw-Hill, 1962

Tsytovich, V. N.: Nonlinear effects in plasma (transl. by M. Hamberger). New York: Plenum Press, 1970

II. Space Physics

Bauer, S. J.: Physics of planetary ionospheres. Berlin-Heidelberg-New York: Springer, 1973

Dyer, E. R. (ed.): Solar terrestrial physics/1970. Dordrecht, Holland: Reidel Publishing, 1972

Helliwell, R. A.: Whistlers and related ionospheric phenomena. Stanford, California: Stanford University Press, 1965

Hess, W. N., Mead, G. D. (ed.): Introduction to space science. New York: Gordon & Breach, 1968

Hundhausen, A. J.: Coronal expansion and solar wind. Berlin-Heidelberg-New York: Springer, 1972

Jacobs, J. A.: Geomagnetic micropulsations. Berlin-Heidelberg-New York: Springer, 1970

McCormac, B. M. (ed.): Particles and fields in the magnetosphere. Berlin-Heidelberg-New York: Springer, 1970

Roederer, J.G.: Dynamics of geomagnetically trapped radiation. Berlin-Heidelberg-New York: Springer, 1970
Schulz, M., Lanzerotti, L.J.: Particle diffusion in the radiation belts. Berlin-Heidelberg-New York: Springer, 1974
Simon, A., Thompson, W.B. (ed.): Advance in plasma physics, vol. I and II. New York: Interscience Publishers, 1968
Yeh, K.C., Liu, C.H.: Theory of ionospheric waves. New York: Academic Press, 1972

III. Plasma Instabilities

Auer, G., Cap, F., Floriani, D., Gratzl, H.: Literature survey on plasma instabilities, vol. I and II. U.S. Dept. of Commerce, National Technical Information Service, AD 722-085 (1970) and AD 738, 218 (1971)
Briggs, R.J.: Electron stream interaction with plasmas. Cambridge, Massachusetts: M.I.T. Press, 1964
Mikhailovskii, A.B.: Theory of plasma instabilities. Engl. Transl. Air Force Systems Command, Foreign Technology Division, FTD-HC-23-0735-72

List of Symbols

$A_{\boldsymbol{k}}$	normalized Fourier amplitude with the wave vector $\boldsymbol{k}$; Eq. (4.85)
$\boldsymbol{B}$	magnetic field vector
c	speed of light in vacuum; 2.998×10^8 m/sec
c_s	ion sound speed $v_{Te}(m_e/m_i)^{1/2}$
D	diffusion constant
$D(\omega, \boldsymbol{k})$	dispersion relation; $D(\omega, \boldsymbol{k}) = 0$
$\boldsymbol{E}$	electric field vector
$\boldsymbol{E}_{\boldsymbol{k}}$	Fourier amplitude of an electric field vector with the wave vector $\boldsymbol{k}$
$\boldsymbol{E}(\boldsymbol{k})$	spatial Fourier transform of an electric field vector
e; e	electron charge; 1.602×10^{-19} coulomb; base of natural logarithm
$\boldsymbol{F}$	force
$f(\boldsymbol{x}, \boldsymbol{v}, t)$	distribution function in $\boldsymbol{v}$-$\boldsymbol{x}$ phase space
$\hbar$	Planck's constant/2π; 6.626×10^{-34} joule · sec/2π
Im	imaginary part
I_n	modified Bessel function of the first kind
$I(\boldsymbol{k})$	spectral density; e.g. $\lim_{V \to \infty} \frac{1}{V} \frac{1}{(2\pi)^3} \lvert E(\boldsymbol{k}) \rvert^2$
i	$\sqrt{-1}$
$\boldsymbol{J}$	current density
$J(\boldsymbol{k})$	action spectral density; $I(\boldsymbol{k})/\omega_{\boldsymbol{k}}$
J_n	Bessel function of the first kind
K_n	modified Bessel function of the second kind
$\boldsymbol{k}$	wave vector
k_D	Debye wave number; ω_p/v_T
L	invariant drift shell parameter; at the equator = radial distance/R_E
m_e	electron mass; 9.107×10^{-31} kg
m_i	ion mass
$N_{\boldsymbol{k}}$	number density of wave quanta with the wave vector $\boldsymbol{k}$: $W_{\boldsymbol{k}}/\hbar\omega_{\boldsymbol{k}}$
n; n	number density of particles; harmonic number

n_c	number density of cold particles
n_h	number density of hot particles
$O(x)$	order of x in magnitude
P	principal value
p	pressure
q	charge of a particle; e for proton, $-e$ for electron
r	radius
R_E	earth radius
Re	real part
$S_{\boldsymbol{k}}$	sign of energy of a wave with the wave vector $\boldsymbol{k}$; sign $\partial\varepsilon/\partial\omega_{\boldsymbol{k}} = \text{sign}\, W_{\boldsymbol{k}}/\omega_{\boldsymbol{k}}$
T	temperature (energy unit)
t	time
u	speed
V	$V_{\boldsymbol{k}}(\boldsymbol{k}', \boldsymbol{k}'')$; coupling coefficient among waves, Eq. (4.89)
$\boldsymbol{v}$	velocity
$\boldsymbol{v}_g$	group velocity, $\partial\omega/\partial\boldsymbol{k}$
$\boldsymbol{v}_p$	phase velocity, $\omega/\boldsymbol{k}$
$v_\parallel$	parallel speed in cylindrical coordinates with $\boldsymbol{z}\parallel\boldsymbol{B}_0$; v_z
v_A	Alfvén speed; $c\,\omega_{ci}/\omega_{pi}$
v_D	drift speed
v_d	diamagnetic drift speed
v_T	thermal speed; $(T/m)^{1/2}$
$v_\perp$	perpendicular speed in cylindrical coordinates with $\boldsymbol{z}\parallel\boldsymbol{B}_0$; $(v_x^2+v_y^2)^{1/2}$
w	kinetic energy; $m(v_\perp^2+v_\parallel^2)/2$
$W_{\boldsymbol{k}}$	wave energy density with the wave vector $\boldsymbol{k}$
W_T	field energy density at the thermal equilibrium; T/λ_D^3
$\boldsymbol{x}$	position vector
x	coordinate
y	coordinate
Z	plasma dispersion function, Eq. (2.31)
z	coordinate
β	pressure ratio of plasma to magnetic field; $n_0 T/(B_0^2/2\mu_0)$
$\gamma_{\boldsymbol{k}}$	growth rate of a wave with the wave vector $\boldsymbol{k}$
$\boldsymbol{\nabla}$	$\boldsymbol{e}_x\dfrac{\partial}{\partial x}+\boldsymbol{e}_y\dfrac{\partial}{\partial y}+\boldsymbol{e}_z\dfrac{\partial}{\partial z}$
δ	$\delta^3=-\eta/2$, η is the density ratio of beam to plasma
$\delta(x)$	delta function
δ_{ij}	Kronecker symbol; $=1$ $(i=j)$, $=0$ $(i\neq j)$
ε_0	permitivity of free space; 8.854×10^{-12} farad/m
ε, $\varepsilon(\omega, \boldsymbol{k})$	equivalent dielectric constant
λ	wavelength

λ	$k^2 v_{T\perp}^2/\omega_c^2$
λ_D	Debye wavelength; v_T/ω_p
$\nu_{\alpha\beta}$	collision frequency of particle α to particle β
μ	mobility; $e/\nu m$
μ	adiabatic invariant; $mv_\perp^2/(2B_0)$
μ_0	permeability of free space; $4\pi \times 10^{-7}$ henry/m
η	resistivity
σ	conductivity
σ	collision cross section
$\sigma(\omega, \boldsymbol{k})$	equivalent conductivity; $-i\omega\varepsilon_0\varepsilon(\omega, \boldsymbol{k})$
κ	measure of density gradient; $-\nabla(\ln n_0)$
ϱ	charge density
ϱ	Larmor radius
τ	mean free time
τ	reduced time
φ	electrostatic potential
Φ	amplitude of a modulated wave
Ψ	magnetic scaler potential
$\omega_{\boldsymbol{k}}$	solution for ω for a given value of $\boldsymbol{k}$ of a linear dispersion relation
ω_p	plasma (angular) frequency; $(e^2 n_0/\varepsilon_0 m)^{1/2}$
ω_c	cyclotron (angular) frequency; qB_0/m
ω_b	bounce (angular) frequency
ω^*	drift wave (angular) frequency; $k_\perp v_T^2 \nabla(\ln n_0)/\omega_c$
$\boldsymbol{\xi}$	reduced spatial coordinate vector
$\boldsymbol{\xi}$	plasma displacement; $\partial\boldsymbol{\xi}/\partial t = \boldsymbol{v}$
$\langle\ \rangle$	ensemble average
Superscript *	complex conjugate
Subscript e	electron
Subscript i	ion (proton)
Subscript i	imaginary part
Subscript p	plasma
Subscript r	real part
Subscript s	stream
Subscript 0	unperturbed (dc) quantities
Subscript 1	perturbed (ac) quantities
Subscript $\perp$	perpendicular to the magnetic field
Subscript $\parallel$	parallel to the magnetic field

Subject Index